MILITARY AIR TRANSPORT OPERATIONS

Brassey's Air Power: Aircraft,
Weapons Systems and Technology Series

VOLUME 6

Brassey's Air Power:
Aircraft, Weapons Systems and Technology Series

General Editor: AIR VICE MARSHAL R. A. MASON, CB, CBE, MA, RAF

This new series, consisting of eleven volumes, is aimed at the international officer cadet or junior officer level and is appropriate to the student, young professional and interested amateur who seek sound basic knowledge of the technology of air forces. Each volume, written by an acknowledged expert, identifies the responsibilities and technical requirements of its subject and illustrates it with British, American, Russian, major European and Third World examples drawn from recent history and current events. The series is similar in approach and presentation to the highly successful Brassey's Battlefield Weapons Systems and Technology Series, and each volume, excluding the first, has self-test questions.

Brassey's Titles of Related Interest

MILITARY AIR
TRANSPORT
OPERATIONS

Group Captain Keith Chapman MPhil, BA, RAF

BRASSEY'S (UK)
(a member of the Maxwell Pergamon Publishing Corporation plc)

LONDON · OXFORD · WASHINGTON · NEW YORK
BEIJING · FRANKFURT · SÃO PAULO · SYDNEY · TOKYO · TORONTO

U.K.	Brassey's (UK) Ltd.,
(Editorial)	24 Gray's Inn Road, London WC1X 8HR, England
(Orders)	Brassey's (UK) Ltd.,
	Headington Hill Hall, Oxford OX3 0BW, England
U.S.A.	Brassey's (US) Inc.,
(Editorial)	8000 Westpark Drive, Fourth Floor, McLean,
	Virginia 22102, U.S.A.
(Orders)	Pergamon Press, Inc., Maxwell House, Fairview Park,
	Elmsford, New York 10523, U.S.A.
PEOPLE'S REPUBLIC	Pergamon Press, Room 4037, Qianmen Hotel, Beijing,
OF CHINA	People's Republic of China
FEDERAL REPUBLIC	Pergamon Press GmbH, Hammerweg 6,
OF GERMANY	D-6242 Kronberg, Federal Republic of Germany
BRAZIL	Pergamon Editora Ltda, Rua Eça de Queiros, 346,
	CEP 04011, Paraiso, São Paulo, Brazil
AUSTRALIA	Brassey's (Australia) Pty. Ltd.,
	P.O. Box 544, Potts Point, N.S.W. 2011, Australia
JAPAN	Pergamon Press, 5th Floor, Matsuoka Central
	Building, 1-7-1 Nishishinjuku, Shinjuku-ku, Tokyo 160,
	Japan
CANADA	Pergamon Press Canada Ltd., Suite No. 271,
	253 College Street, Toronto, Ontario, Canada M5T 1R5

Copyright © 1989 Brassey's (UK) Ltd.

First edition 1989

Library of Congress Cataloging in Publication Data
Chapman, Keith.
Military air transport operations/Keith Chapman. — 1st ed.
p. cm. — (Brassey's air power: v. 6)
1. Airlift, Military. I. Title. II. Series.
UC330.C47 1989 358.4'4 — dc 19 88-38897

British Library Cataloguing in Publication Data
Chapman, Keith.
Military air transport operations —
(Brassey's air power: aircraft, weapons systems,
and technology series; V.6)
1. Air transport by military aircraft
I. Title
358.4'4

ISBN 0-08-034749-5 Hardcover
ISBN 0-08-036255-9 Flexicover

The front cover illustration is an artist's impression of McDonnell Douglas C-17 airlifters operating at an austere, forward airstrip in the 1990s. The C-17 is due to enter service with the USAF's Military Airlift Command in 1992. (By courtesy of McDonnell Douglas)

Printed in Great Britain by BPCC Wheatons Ltd., Exeter

About the Author

Group Captain K. Chapman M Phil, BA, RAF entered the Royal Air Force in 1960. His flying career, which culminated in a tour on Hercules aircraft at RAF Lyneham, as Officer Commanding No. 24 Squadron, has been spent exclusively in tactical transport operations. Other appointments have included tours as Personal Air Secretary to the Minister for the Royal Air Force; RAF member of the directing staff at the Army Staff College, Camberley; and Deputy Director of Air Transport Operations in the Ministry of Defence in London. Granted a sabbatical year at Magdalene College, Cambridge in 1980/81, he was the first RAF officer to receive that University's degree of Master of Philosophy in International Relations. He is currently serving on the staff of the Supreme Headquarters Allied Powers Europe.

Preface

By virtue of their versatility, reach and quick reflexes, air transport forces have an important part to play in the exertion of political and military power, both at the strategic and tactical levels. With that firmly in mind, I have sought to examine the various categories of fixed and rotary wing operations in terms of principles, requirements, planning factors and techniques. In order to put flesh on these bones, I have also included details of some of the main types of aircraft involved. However, I have excluded coverage of the VIP and communications role since these are outside the scope of a book concerned essentially with the operational applications of airlift forces.

I am very grateful to everyone who has helped me in the preparation of this book, especially my wife Lyn who worked long hours to turn my hand-written notes into a highly professional typescript. I am also indebted to various aerospace companies in Europe and the USA who provided me with valuable technical information and illustrative material. However, reference to a particular aircraft or item of equipment does not necessarily signify an endorsement of its quality or operational value.

Finally, I must emphasise that the facts and opinions expressed in this book are entirely the author's responsibility and imply no endorsement either by the Ministry of Defence of the United Kingdom or by any other agency.

K.C.

Contents

List of Figures

List of Plates

1

Introduction

THE AIR POWER CONTEXT

Considering the sophisticated inventories and impressive capabilities of modern air forces, it is easy to forget that military aviation is, in historical terms, still very much in its infancy. Disregarding earlier attempts to use balloons and airships as instruments of war, the application of air power dates back only some 75 years. However, in that relatively short space of time, air power has established itself as an extremely effective means of exerting military force, one of its chief advantages being its versatility. Thus an air force is not only able to deliver firepower directly by employing air-to-surface and air-to-air weapons, but can also be used to bring force to bear indirectly by, for example, mounting airlift operations to deploy combat units at high speed over substantial distances. But if air power in general is still relatively new, then air transport operations—which have rapidly evolved to occupy a pivotal position in the force projection equation—are an even more recent development; for although rudimentary transports were used on a limited scale in support of air policing and other operations during the inter-war years,[1] the wider exploitation of military airlift dates only from 1939. Until that time, the deployment and logistic support of military forces was undertaken almost exclusively by surface transport systems.

The traditional projection of power by land and sea was by definition a laborious and protracted process, often involving a degree of vulnerability *en route* that threatened the attrition or destruction of a force before it could even reach its objective. Indeed, the history of warfare is riddled with occasions when surface-deployed forces either arrived too weakened, too late or not at all. Similarly, the employment of 'gunboat diplomacy' in this and earlier centuries often failed to exert the required leverage for want of being able to get the 'gunboat' into position at exactly the right moment. In more recent times, it is pertinent to note that the Arab Israeli war of 1973—admittedly a brief affair—was over by the time that the first sea-delivered supplies arrived in Israel from the USA. Meanwhile, the USSR had airlifted some 15,000 tons of war *matériel* to Egypt and Syria, and about 27,000 tons had been flown into Israel by a combination of US/Israeli civilian and military air transport assets.

Although technological advances during this century have made surface deployment a speedier and more efficient process, the movement of forces by rail, road and sea is still in many circumstances either too slow, too restricted by geographical constraints, or too susceptible to hostile interception. It was the growing perception of such limitations, coupled with an increasing awareness of what airlift could offer in terms of speed, reach and capacity, that led both the Allies and their opponents (especially the Luftwaffe) to give much greater priority than hitherto to the build-up

PLATE 1.1.   C-5 Galaxy.

of their respective air transport forces from 1939 onwards. From modest beginnings, these forces were rapidly expanded to become a significant component of the air order of battle. For example, the USA's Air Transport Command, formed only in June 1942, was by July 1945 carrying 275,000 passengers and 100,000 tons of cargo per month on a worldwide network of routes using a fleet of 3,700 aircraft.[2] Whilst it would be an exaggeration to claim that airlift operations were crucial to the overall outcome of the Second World War, there is no doubt that they played a major and in some cases a decisive role in certain specific scenarios.

This was especially true in the Far East, where a combination of vast distances, difficult terrain and enemy dispositions made surface movement an extremely arduous if not impossible proposition. The Allied campaign in Burma during the latter half of the war relied particularly upon the adroit exploitation of airlift assets. Operating under the umbrella of virtual air superiority which the British and Americans had jointly managed to establish over the Japanese by late 1943, General Slim (Commander of the British 14th Army) used transport aircraft not only to deploy and redeploy entire divisions, but also to sustain them in situations where surface supply would have been immensely difficult if not out of the question. For example, a force of 155,000 men was supplied wholly by air for nearly 3 months—involving an average delivery of 250 tons of food and ammunition per day—during the Japanese siege of the twin British positions at Kohima and Imphal in 1944. Following the lifting of that siege, Slim regained the initiative and, during its subsequent advance from the Indo/Burmese border to ultimate victory in Burma in May 1945, his 14th Army—a force of 300,000 men—received no less than 90% of its total supplies by air. Transport squadrons also played a leading role in the two 'Chindit' expeditions into Japanese-held territory in 1943 and 1944. The second of these operations entailed the delivery by RAF and American aircraft of 9,000 men, 1,360 mules and ponies, and 227 tons of supplies into a number of improvised airstrips carved out of the jungle some 150 miles behind enemy lines. Most of this force was landed at an airstrip nicknamed 'Broadway'. Having witnessed this impressive operation, Air Marshal Baldwin (Commander of the Tactical Air Force of South East Asia Command) was later to write:

> 'Nobody has seen a transport operation until he has stood at Broadway under the light of a Burma moon and watched Dakotas coming in and taking off in opposite directions on a single strip at the rate of one take-off or one landing every three minutes.'[3]

POST-WAR EVOLUTION OF AIR TRANSPORT OPERATIONS

Although air transport operations had certainly burgeoned into a major facet of air power by 1945, ground forces continued to move and to be supplied predominantly by land and sea, regardless of the distances involved, well into the 1950s. This was partly because military doctrine still leaned heavily towards surface deployment and partly because airlift resources were still relatively limited *vis-à-vis* the scale of the overall task. Indeed, by virtue of their sheer size and bulk, many cargoes were simply not air-transportable—a practical constraint which even today can present serious problems for the airlift planner. Nevertheless, just as the extensive use of fighter, bomber and maritime aircraft in the Second World War demonstrated the critical importance of being able to bring direct force to bear in and from the 'third dimension', so the advent of transport operations provided the first real opportunity for airlift forces to unveil their potential for the application of both direct and indirect force. They did this by repeatedly airlifting troops, equipment and logistic supplies over terrain and within timescales that would formerly have been unattainable.

One event above all others in the immediate post-war era was to provide an opportunity for air transport forces to demonstrate their politico-military worth. This occurred in 1948 when the USSR suddenly imposed a blockade of all surface access routes to West Berlin. Determined that the two million citizens of that city should not be abandoned to their fate, yet anxious not to provoke the direct military confrontation that might easily have resulted from despatching an armed convoy with orders to force a way through by road, the Allied leaders responded by launching an airlift to the beleaguered Western sectors of Berlin. To the surprise of many—not least the Soviet leadership—the collective airlift assets of the USAF and RAF, with some assistance from French military and British civilian sources, proved more than equal to this formidable challenge. Operating aircraft which were much less sophisticated and capable than today's transports, the Allies airlifted more than two and a quarter million tons of food, coal, oil and other supplies into West Berlin during the 15 months from June 1948 to September 1949.[4]

This impressive achievement not only demonstrated in practical terms that the Allies could sustain a major airlift for an indefinite period but, by persuading the Kremlin that its blockade could not succeed, was also largely responsible for permitting the peaceful settlement of a dispute which might otherwise have sparked off another major war in Europe. Perhaps more significant, the Berlin Airlift showed beyond doubt that air transport forces could be a powerful factor in the application of political as well as military pressure.

PLATE 1.2.   Scene at RAF Gatow during the Berlin Airlift.

During the four decades that have followed, all major powers—and many lesser ones—have increasingly come to recognise that a sovereign state must, in the final analysis, be ready and able to exert force or at least the threat of force in defence of its vital interests. These cover a broad spectrum and vary widely from state to state but include such common imperatives as resistance to internal subversion, defence of national territory and maintenance of access to international trade routes. If, when such interests are jeopardised, sufficient forces are not already in place, they must be moved promptly and efficiently to where they are needed. Almost always, time will be of the essence; and in most cases only airlift will ensure the necessary speed of reaction. The implications of this military fact of life are self-evident, and most states now accept that an airlift capability is as crucial as any other component of air power to the overall balance, effectiveness and credibility of their respective defence postures. This is reflected in the size and shape of air forces around the world: Table 1.1 gives some measure of the resources currently involved.

AIRLIFT AS AN INSTRUMENT OF FORCE PROJECTION

Investment in air transport forces on this scale has occurred only because the states concerned are well aware that neither military forces in general nor army and air combat units in particular can realise their full potential unless they can be rapidly brought to bear wherever they are needed. This is not to suggest that forces can be expeditiously deployed only by air. In some cases, perhaps involving deployments of just a few hundred miles, it may be almost as quick and certainly more cost-effective to deploy by truck or train if suitable roads or railways are available. Moreover,

TABLE 1.1
Airlift assets of some major air forces

Country	Tanker transports	Strategic airlifters	Tactical airlifters
USA[1]	60	415	728
USSR[2]	–	363	260
People's Republic of China	–	28	161
UK	9[3]	13	61
Federal Republic of Germany	–	4	75
France	–	5	72
Italy	–	–	42
Canada	–	5	28
Israel	–	5	22

Notes:
1. Includes Air Force Reserve and Air National Guard assets.
2. Figures quoted indicate aircraft operated by Military Transport Aviation Command – otherwise known as *Voyenno-Transportnaya Aviatsiya (VTA)*.
3. Planned figure; all nine aircraft not yet in service.

PLATE 1.3.   AN-124 Condor.

PLATE 1.4.   C-160 Transall.

airlift can never compete in terms of payload with the massive capacity of a good railway system or fleet of cargo ships. On the other hand, the fact remains that the projection and employment of military force can seldom if ever be satisfactorily achieved without recourse to at least some airlift.

For example, a fighter squadron cannot operate effectively away from its home base without groundcrew, specialist equipment and weapons; most if not all of this logistic support must be airlifted if the inherent speed and flexibility of the fighters themselves are not to be seriously degraded. Furthermore, depending on the characteristics of the combat aircraft in question, the distance to be flown, and the availability of staging airfields and overflight clearances, fast jet units may be unable to redeploy to where they are needed unless air-to-air refuelling (AAR) is provided. As explained in more detail in Chapter 3, the development of tanker transport aircraft to perform airlift and AAR tasks simultaneously serves to enhance both the 'force multiplier' and 'force extender' concepts, by conferring greater flexibility on deployment options and increasing the reach of combat squadrons.

The fundamental importance of airlift—both fixed and rotary wing—to the employment of ground forces cannot be emphasised too strongly. In today's volatile international arena, where actual or potential crises and conflicts are ever liable to develop with little or no warning, speed of reaction is absolutely vital. Hence armies achieve and retain full credibility only if they are able—and seen to be able—to deploy rapidly from their peacetime bases to wherever they may be needed. Ground forces which cannot call upon at least some dedicated airlift lack reflexes and mobility, and as a result may be impotent to intervene effectively in situations where they might otherwise (with the benefit of air mobility) be able to play a decisive role.

That said, it would be unrealistic to suggest that military forces could or should be deployed, resupplied and redeployed exclusively by air; this is a highly expensive option which few states could justify or afford. Even a relatively small-scale operation, especially if it includes the deployment of armoured vehicles, artillery and helicopters, cannot be mounted entirely by air without substantial airlift assets. Hence most states, including those whose resources are more or less in line with their national commitments and international aspirations, settle for a more limited air transport capability, accepting any disadvantages that this policy may entail. For example, while France has for some time maintained a sizeable rapid deployment force of some 47,000 personnel, it still lacks the strategic assets to airlift more than a small proportion of these troops and their equipment to theatres outside Europe.[5] Partly as a result of this lack of strategic air mobility, France continues to maintain large and expensive garrisons in distant locations which could otherwise be run at much lower strengths, relying on rapid reinforcement by air when and if required—as is now the case with the British garrison in the Falkland Islands.

At this point, it is necessary to clarify the distinction between rapid reinforcement (which, as the term suggests, entails the deployment of forces by air to strengthen an existing base or garrison) and the wider concept of using airlift to intervene in areas where there was previously no military presence. While the ability to undertake rapid reinforcement operations offers a number of political, operational and economic advantages, especially if heavy equipment can be permanently stockpiled in the relevant locations, the capacity to insert forces by air whenever and wherever they

PLATE 1.5.   KC-10 Extender.

may be required is arguably much more important. That is why the USA continues to build up its airlift force in line with the development of its concept for the Central Command (Cent Com) which was established in 1983 as the successor to the Rapid Deployment Joint Task Force.[6] Based in the USA, but with an area of responsibility encompassing 19 countries in South-West Asia, the Middle East and North-East Africa, HQ Cent Com is some 7,000 miles by air (equivalent to 15 hours on a flight-refuelled non-stop C-5 Galaxy) from the Persian Gulf. The distance by sea is about 8,000 miles via the Suez Canal or 12,000 miles around the Cape of Good Hope. As the curent Commander in Chief of Cent Com recently emphasised, his major challenge lies in moving his forces over substantial distances in short timescales. To quote his own words, 'the key to doing this is rapid force projection, achieved by using a combination of airlift, sealift and pre-positioning'.[7]

Despite their massive military budgets and their large air transport fleets, neither the USSR nor the USA has sufficient airlift assets to deploy a force such as Cent Com, or satisfy its insatiable appetite for logistic support once in place, entirely by air. Nevertheless, it is clear that both Superpowers are firmly committed to the proposition that early deployment of even a limited force can often defuse or stabilise a potentially dangerous or deteriorating situation, while simultaneously sending a strong political signal to any third parties concerned. A number of recent examples serve to illustrate this basic principle, including the Soviet airlift into Afghanistan in 1978 and the USA's airlift of additional forces (albeit temporarily) into Panama in 1988.

Although less powerful states have far fewer resources than either Superpower, the same axiom holds good—namely that the availability of enough airlift to deploy some degree of military force will confer a measure of potential out of all proportion to the expenditure or opportunity-costs involved. NATO's Allied Command Europe Mobile Force (better known as the ACE Mobile Force or more simply as the AMF) is an excellent illustration of this concept. A multinational formation comprising several squadrons of combat aircraft and a brigade-sized ground component of some 5,500 men and 1,500 vehicles, the AMF is a lightly armed force which is specifically designed for immediate reaction. Its three main tasks are to reinforce deterrence in a threatened area; to demonstrate NATO resolve and solidarity; and to work with host nation forces to strengthen local defences. In order to be in place in time to achieve these objectives, the AMF must clearly move with great speed to its deployment areas from its dispersed locations on both sides of the Atlantic (see Figure 1.1). This means that the AMF must rely heavily upon air transport.

As a further illustration of the versatility of air transport operations, it is important to remember that airlifters can make a crucial contribution to a military campaign even if they are unable to land in the operational area in question. Instances are not difficult to find in recent military history, the German airborne assault on Crete in 1941 and the Allied landings behind the Normandy beach-heads in 1944 being two striking examples of the delivery of troops and equipment into battle by parachute. Some 40 years later, in 1982, Hercules aircraft of the Royal Air Force were called upon to play a vital role in the South Atlantic conflict well before any airfield on the Falkland Islands had been recaptured by the British Task Force. Operating from Ascension Island (see Figure 1.2) and using AAR in order to undertake round trips which regularly exceeded 24 flying hours, these aircraft were tasked to air-drop

PLATE 1.6. RAF C-130 arriving at Erzurum air base in Turkey with troops and equipment for AMF Exercise AURORA EXPRESS, June 1987.

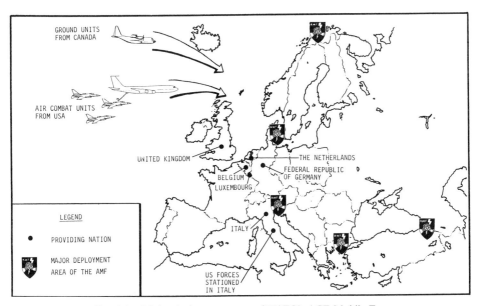

FIG. 1.1. Major deployment areas of NATO's ACE Mobile Force.

PLATE 1.7.   US infantry deplaning from C-130 after landing at forward airstrip.

essential items of equipment—and even a few key personnel—to naval and army units in the combat zone. Without the reach and speed of such operations, which allowed equipment to be delivered to the Falklands area from the UK in under 24 hours, there is no doubt that the Task Force's mission could have been seriously hampered or, worse still, jeopardised. Nevertheless, one should not infer from such examples that air-drop operations are universally feasible and successful; on the contrary, they are heavily dependent upon particular circumstances, not least the existence or otherwise of a favourable air situation (see Chapter 6).

CIVILIAN APPLICATIONS

While the provision of air transport forces must of necessity be geared primarily to military needs and applications, the possession of such assets does of course permit a state to employ them on non-military tasks when it so chooses. Typically, such tasks include charter to other government agencies (eg, for the carriage of bulky cargo which is beyond the capacity of available civilian aircraft); carrying Heads of State and other VIPs on both domestic and international flights; assistance with projects in aid of the civilian authorities (eg, emergency airlift of police reinforcements or fire-fighting teams); and participation in humanitarian operations.

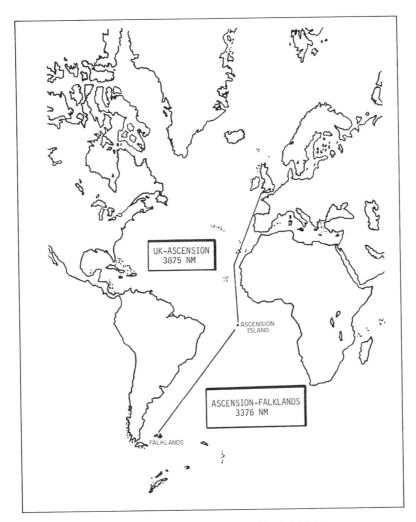

FIG. 1.2. Route flown by RAF on Falklands Airbridge.

A good example of this last occurred in 1984/85 when the governments of the UK, Belgium, Federal Republic of Germany, German Democratic Republic, Italy, USSR and Poland all provided air transport detachments to assist with famine relief operations in Ethiopia. Over a period of 409 consecutive days, the Royal Air Force detachment of two Hercules alone delivered a total of 32,160 tons of grain and other urgently needed supplies, 14,380 tons of which were air-dropped into remote areas where no suitable airstrips were available.

The benefit of such humanitarian operations are manifold. First, they make an invaluable contribution to the relief of human suffering. In Ethiopia, where a combination of very difficult terrain and poor surface communications made air transport the best and, in some cases, the only means of distributing food to drought-stricken areas, there is no doubt that the multinational airlift operation saved many thousands of lives. Second, there are usually direct benefits for the participants as

PLATE 1.8. RAF C-130 Hercules in Ethiopia.

well. Crews who served in Ethiopia gained unique exposure to tactical operations in a hot and mountainous environment, thereby enhancing their wider skills and experience. Finally, governments themselves often gain a useful spin-off, in terms of international prestige and strengthened domestic support for defence spending, from employing transport forces on humanitarian operations, while the general public is always reassured to see that expensive military assets can sometimes be harnessed to non-defence tasks for the wider benefit of mankind. In short, transport operations in support of the civil authorities or humanitarian projects confer both political and military advantages, which jointly do much to enhance the public perception of the armed forces in general and which, incidentally, help to reinforce the need for a military airlift capability.

IMPORTANCE OF AIRLIFT TO OVERALL MILITARY POSTURE

As emphasised earlier in this chapter, the success or failure of a military operation will often be determined by the speed with which the required forces can be moved into position. It is this factor above all others which underlines the importance of an airlift capability. While larger quantities of equipment and supplies can usually be deployed more economically by sea, rail and road, the relatively slow speed of surface movement systems cannot compete with airlift when the rapid insertion of force is required. Thus, for example, the spectacularly successful Israeli operation at Entebbe in 1976 and the less dramatic but equally decisive Franco/Belgian operation at

Kolwezi in 1978—each of which hinged crucially upon reach, speed and surprise—could only have been carried out by airlifted forces. In short, the operational posture, flexibility and capability of a given force are significantly conditioned by the extent to which its key components can be airlifted. Hence transport operations are not only an important facet of air power in their own right but also have a critical role to play in support of overall military strategy and capability.

AIR TRANSPORT ROLES

Having considered the significance of an airlift capability for the exploitation of a state's wider military potential, it is necessary to examine in more detail the various roles of air transport operations and the principles whereby they are conducted. In essence, the primary tasks of transport forces are to provide airlift and logistic support for other military units both between and within theatres of operation, including the immediate vicinity of the battlefield. Thus air transport tasks fall mainly into the following two categories:

Strategic Operations
The airlift of personnel and equipment *between* theatres (eg, between the USA and Europe—see Figure 1.3) including scheduled, specially mounted and aeromedical evacuation flights.

Tactical Operations
The airlift of personnel and equipment *within* a theatre (eg, southern flank of NATO—see Figure 1.4). Tactical tasks include:

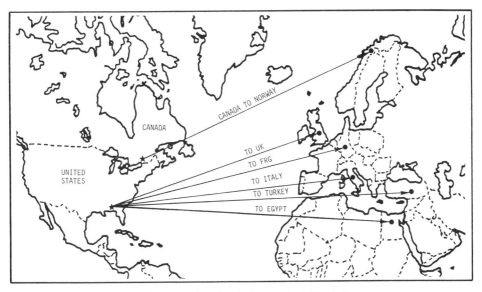

Fig. 1.3. Examples of typical strategic airlift missions.

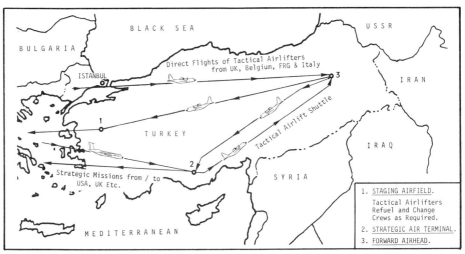

FIG. 1.4. Example of a theatre airlift operation on NATO's south-eastern flank.

- Air-Landed Missions—delivery by fixed wing aircraft into prepared or semi-prepared airfields.
- Air-Dropped Missions—delivery by parachute from fixed-wing aircraft either because the operational situation demands an airborne assault of troops and air-drop of equipment, or because no suitable airfield is available for air-landed missions.
- Heliborne Missions—delivery by helicopter either directly into battle (eg, for a *coup de main* operation) or to other locations designated by the ground force commander.
- Special Forces Missions—insertion of special forces either by air-landing, air-dropping or helicopter.
- Aeromedical Evacuation Missions—evacuation of casualties by helicopter and/or fixed-wing aircraft from forward to rear areas.

PRINCIPLES OF AIR TRANSPORT OPERATIONS

Air transport is a valuable and expensive resource for which demand invariably exceeds supply. Accordingly, certain principles must be observed if airlift assets are to be exploited in the most effective and economical manner. The most important of these is the need for the direction of air transport forces to be undertaken at the highest possible level in order to ensure that conflicting demands and priorities are properly assessed and resolved. This is especially important where strategic and the larger tactical airlifters are concerned, since the correct employment of such aircraft could be crucial to the success of any major deployment or reinforcement. In the case of support helicopters, operational command and control may need to be delegated to lower levels for specific operations or when circumstances so dictate. This aspect of helicopter tasking is examined in more depth in Chapter 7.

An associated and equally important principle hinges upon the need to ensure maximum economy in the use of airlift. While transport aircraft frequently offer the best if not the only means of enabling the rapid deployment of combat forces, they

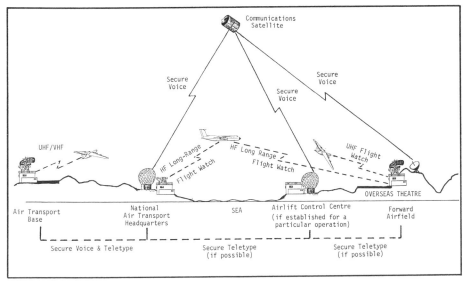

FIG. 1.5.   Example of modern communications network used for command and
control of transport operations.

are costly to operate and must therefore be used as efficiently as possible. With that
in mind, the cardinal rule is that surface transport systems (such as sea, rail and
road) should be used in preference to, or in combination with, airlift whenever this
is feasible and operationally acceptable. Even if airlift is used exclusively for the initial
phase of a deployment, surface lines of communication should be established as soon
as possible thereafter in order to husband airlift for other tasks which can be
undertaken only by air. Economical use of air transport calls for good management
and co-ordination at all levels in the tasking process, which in turn depend heavily
upon effective communications. Without the necessary investment in state-of-the-art
communications systems (such as SATCOM secure voice and data links) command
and control of transport operations could be seriously impaired.

The third important principle is that airlift operations should be planned *jointly*
from the outset to ensure that the requirements and interests of both 'customer' and
'operator' are fully considered. For example, an army unit will want its troops and
equipment to be delivered to the destination airfield, dropping zone or helicopter
landing site in a designated sequence within a precisely defined timescale. These
specifications must be balanced against the operator's need to select timings, routes
and flight profiles which take account of such factors as overflight clearances,
availability of staging posts and threats from hostile forces. Planning would also need
to take account of the prevailing air situation; owing to the marked vulnerability of
transport aircraft to enemy air action, a favourable air situation is an essential
prerequisite for the employment of airlift assets in most operational scenarios.

2

Strategic Operations

As explained in Chapter 1, airlift operations play a crucial role in the projection of military and political power, especially when it is necessary to exert such pressure over substantial distances and at short notice. It follows that the capability to mount strategic transport operations—defined as the airlift of personnel and cargo between theatres of military activity or major geographical areas—is particularly important to any state which aspires to project its influence beyond its borders. This does not mean, of course, that inter-theatre operations are exclusively conducted within the context of power projection. On the contrary, strategic airlift has many wider applications including aeromedical evacuation, the resupply or reinforcement of existing garrisons, support for exercises and humanitarian relief operations. Whatever the mission, the exacting nature of strategic operations—usually involving the airlift of considerable payloads over intercontinental distances—requires aircraft employed in this role to meet an especially stringent set of criteria. These are discussed below.

AIRCRAFT REQUIREMENTS

Most military aircraft require special capabilities and modern strategic airlifters are no exception. They are large, sophisticated and extremely expensive both to buy and operate. For example the USAF is thought to have paid some $150m (at 1985 prices) for each of its 50 C-5B aircraft, a total outlay of $7.5bn even before life cycle costs are added to the equation. Hence it is hardly surprising that the colossal cost of acquiring and operating a large fleet of strategic airlifters is beyond the resources of all but the USA and USSR. Other air forces, including those of the People's Republic of China, UK, France, Canada, Israel and the Federal Republic of Germany have far fewer strategic assets, as shown in Table 1.1. Moreover, most of these aircraft are elderly and hence meet hardly any of the criteria outlined below. On the other hand, most major states can call upon commercial resources in time of crisis to provide modern wide-bodied airliners capable of carrying both passengers and cargo over strategic sectors. As one would expect, such supplementary assets are most readily available in the USA and USSR which can each draw upon substantial military reserves as well as civilian transport aircraft. For example, in addition to significant numbers of Air Force Reserve and Air National Guard airlifters, the Pentagon can also call upon the Civil Reserve Air Fleet (CRAF) which currently comprises over 200 passenger and 100 cargo aircraft from various airlines which have agreed to certain military requirements (such as a strengthened floor) being embodied in return for appropriate federal subsidies. For its part, the Kremlin is not only able to co-opt over a thousand transports currently assigned to air commands other than the

Voyenno-Transportnaya Aviatsiya (*VTA*),[1] but can also direct the massive resources of the state airline Aeroflot, which operates several hundred aircraft (such as the IL-76 and AN-12) capable of immediate adaptation to military tasks.

Aircraft designed for the airlift of troops and military equipment over strategic distances should, as a minimum, satisfy the following basic requirements:

- Large payload in terms of both weight and size, enabling airlift of wide range of military hardware such as artillery, armoured personnel carriers, trucks and small helicopters.
- Ability to carry personnel and/or cargo.
- Long range (at least 2500 nm with maximum payload).
- Rapid onload and offload, with large cargo doors and integral ramps to facilitate carriage of wheeled and tracked vehicles.
- High cruising speed (at least 0.75 Mach).

Ideally, modern strategic airlifters should also be capable of:

- Carrying outsize loads (eg, main battle tanks).
- Operations from austere airfields and unpaved surfaces.
- Air-dropping.
- Receiving fuel in flight.
- Rapid reconfiguration from one role to another (eg, from passenger/cargo to air-dropping).

The need to provide as many of these features as possible presents the designer with a considerable challenge, especially if the aircraft is to be capable of combined strategic/tactical missions such as a trans-oceanic flight followed by an air-drop or by a landing at a semi-prepared airfield. Driven—and constrained—by the criteria listed above, the design of strategic airlifters has generally featured a high wing equipped with lift-enhancing devices, an elevated tail, a low-slung fuselage and high flotation landing gear (ie, a combination of suspension, wheels and tyres which can absorb severe stresses during landing and take-off). Such features are necessary to allow operations from short/unpaved airfields, and to provide a spacious cargo compartment which can be readily loaded and unloaded through large rear doors and an integral ramp. Within this conventional design, however, there is increasing scope for improvement thanks to the availability of lighter-weight yet more durable materials, advanced high-lift systems and more efficient engines. Such developments have paved the way for the new generation of airlifters discussed in Chapter 8. The remainder of this chapter will focus upon the current generation of strategic airlifters some of which, it will be noted, already feature examples of the new technology mentioned above.

MAJOR STRATEGIC AIRCRAFT IN CURRENT OPERATIONAL SERVICE

C-5 Galaxy

The Lockheed C-5 Galaxy is operated exclusively by the USAF's Military Airlift Command (MAC). When it first entered service in 1970, the C-5 was the world's largest and heaviest aircraft, a mantle which has now been assumed by the Antonov AN-124 Condor which is described later in this chapter. There are two variants of the Galaxy, the C-5A and C-5B. MAC acquired 81 C-5As during the period 1970–73 of which 77 remain in operational service today. Acquisition of the C-5B, an updated and improved version of the C-5A, began in December 1985. By the end of 1988, 46 C-5Bs had been delivered to the USAF. A final batch of four aircraft are scheduled for delivery in 1989, making a total of 50 C-5Bs altogether and an overall C-5 fleet of 127 aircraft (included in the USAF total of strategic airlifters in Table 1.1).

The C-5 Galaxy is specifically designed to undertake a wide and demanding range of airlift tasks. By virtue of its ability to operate from relatively short unpaved airfields, and air-drop both paratroops and cargo, the C-5 can be used on quasi-tactical as well as strategic tasks. However, its concept of operations requires it to

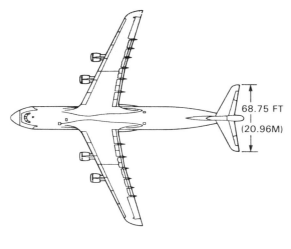

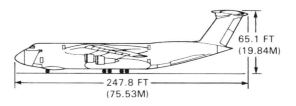

FIG 2.1.   Major dimensions of C-5.

undertake only those tactical missions which have a strategic dimension, such as the air-drop of heavy equipment at a dropping zone in the Middle East after a direct flight from the USA. Similarly, whilst the C-5 is capable of carrying up to 345 troops, it is not normally configured in the full passenger role. Instead, it is used first and foremost as a strategic freighter, this being the mission for which it was primarily procured and which best exploits the ample proportions of its cavernous lower deck. Each version of the C-5 can receive fuel in flight from boom-equipped tankers, thereby allowing the aircraft to achieve ranges constrained only by the limits of crew duty time. The photograph at Plate 2.1, of a C-5A being refuelled over the Grand Canyon by a KC-10, illustrates the extent to which even the KC-10—itself a large wide-bodied aircraft—is dwarfed by the massive Galaxy.

General Description

The C-5 is a high-wing monoplane, swept to 25°, with a 'flying' T-tail. It is powered by four General Electric TF-39-1C turbofan engines mounted under the wing and equipped with cascade-type thrust reversers. The aircraft is divided into three pressurised compartments: an upper level comprising the flight deck and passenger cabin, and a lower cargo hold extending beneath both upper areas.

Flight Deck. The flight deck contains stations for two pilots, a flight engineer and instructor or supernumerary. A rest area with six bunks and seven seats for relief crew members is provided behind the flight deck. There is also a small cabin for use by VIPs or couriers, in the forward section of the upper deck, with eight seats plus toilet and galley.

Passenger Cabin. The main passenger cabin with 75 seats plus additional toilet and galley facilities is in the rear section of the upper deck, aft of the wing. This cabin

PLATE 2.1. C-5A being flight-refuelled by KC-10.

can be used to carry the drivers, crews and operators of vehicles and weapon systems carried in the cargo hold, thereby expediting the offload process at the destination and ensuring that the airlifted equipment is available for operational tasks as soon as possible.

Cargo Hold. The dimensions of the spacious cargo compartment are listed in the Summary of Leading Particulars on page 24 but it should be noted that the maximum width of 19 ft tapers to only 13 ft above the 9.5-ft level. With forward and aft doors which open to expose the full width and height of the cargo hold, integral ramps that extend to ground level, and a kneeling landing gear, the aircraft can be rapidly loaded and unloaded from either end. This drive-on/drive-off arrangement allows vehicles to pass through the entire length of the cargo compartment; hence a deploying unit can drive its vehicles in through the rear when loading and, with the nose raised and forward ramps extended, drive off straight ahead at the destination. Without this valuable drive-through feature, vehicles would either have to be reversed-loaded into the cargo hold before departure to permit a speedy offload, or be laboriously reversed out on arrival. The latter is a time-consuming and operationally questionable procedure, albeit favoured by some air forces equipped with airlifters that have only rear cargo doors.

Landing Gear. The main undercarriage consists of four 6-wheel bogies, each with three pairs of wheels triangularly displaced, while the nose assembly has four wheels mounted in line abreast and steerable to 60° on either side of the aircraft fore and aft axis (illustrated at Plate 2.2).

This high flotation undercarriage, with 28 wheels in all to distribute the considerable weight, makes a major contribution to the C-5's ability to operate from unpaved airfields. It also greatly extends the number of airfields from which the C-5 can operate without exceeding the load classification criteria that are imposed to prevent damage to runways, taxiways and parking areas. For example, at an all-up weight of 665,000 lb, the individual footprint pressure of each of the C-5's 28 wheels is less than the individual footprint pressure of a C-130 weighing 130,000 lb. In practice, this means that, even at high gross weights, the C-5 can normally operate at airfields used by much smaller and lighter aircraft such as the C-130, C-141 and B-707. At medium and light gross weights, C-5 wheel loads are only slightly higher than those of the DC-9 and B-737.

The landing gear also incorporates a hydraulically operated kneeling system which allows the cargo compartment of a parked aircraft to be lowered by about 3 ft to truck-bed level. By reducing the height of the cargo floor in this way, and by decreasing the ramp angles which must be negotiated by vehicles, this feature greatly facilitates and expedites loading and off-loading procedures.

Cargo Capacity. With its built-in roller conveyor system, the cargo compartment can accommodate up to 36 standard 463L pallets, each weighing up to 10,000 lb, arranged side by side between two sets of restraint rails. The conveyor assemblies are normally stowed in the floor and must be turned over manually when required for loading. A winch is also provided.

The floor and ramps can accept tracked vehicles weighing up to 134,200 lb each, and wheeled vehicles with single axle loads of up to 36,000 lb. This load-carrying capacity, combined with full-width access to an exceptionally spacious hold, means that the aircraft can accommodate a wide variety of outsize and exceptionally heavy

PLATE 2.2.   C-5B taking off (landing gear down).

military hardware. For example, it can airlift two M-1 main battle tanks, or eight armoured personnel carriers two abreast, or a large number of jeeps three abreast, or six UH-60A helicopters. In practice, this enables the C-5 to carry virtually every item in the US Army's current inventory of combat equipment.

Air-drop Capability. Special role equipment is needed for air-dropping missions but this can be readily fitted. Heavy equipment—on pallets each weighing up to 40,000 lb—can be air-dropped through the central cargo door in the underside of the rear fuselage; in trials, a Galaxy has dropped four such pallets (a total weight of 80 short tons) in a single pass over a dropping zone. Paratroops can be dropped simultaneously from two doors positioned aft on each side of the cargo hold.

Other Features

Automatic Flight Control System (AFCS). The C-5A has a sophisticated, autopilot-linked AFCS providing:

- Automatic throttle adjustment to control airspeed, Mach number or angle of attack.
- Computerised augmentation of stability in all three axes.

- An active lift distribution control system (ALDCS) that reduces structural stress by automatic aileron deflections in response to gusts and manoeuvres, complemented by automatic elevator movements to counter changes in pitch arising from these aileron deflections.
- A cable servo to assist the pilot to overcome the control breakout force.
- An aircraft attitude system giving optimum pitch commands during rotation, take-off, climb-out and overshoot.

Malfunction Detection Analysis and Recording System (MADAR). The C-5A was equipped from the outset with an on-board computer to monitor—either in flight or on the ground—more than 800 test points in the various systems and sub-systems of the aircraft. Designed (as its name suggests) to detect, identify and record malfunctions, MADAR was a revolutionary feature when it appeared in 1970. With the benefit of new micro-chip technology, the system has been upgraded over the years and an advanced version—designated MADAR II—is installed in the C-5B.

Flying Controls and Lift Devices. The C-5 has conventional ailerons, a rudder divided into two sections and an elevator split on each side. Pitch is trimmed by varying the incidence of the 'flying' tailplane. Aileron control of roll is augmented by the differential use of the outer five (of nine) spoilers on each wing; the inner four spoilers are used, albeit together with the other five, only as lift dumpers. High lift is provided by Fowler flaps and leading-edge slats, which are interconnected so that the slats are fully extended when the flaps reach the 40% down position, and begin to retract automatically as the flaps are brought in through that position. These various lift devices make a major contribution to the Galaxy's ability to operate from relatively short airfields. For example, at a weight of 600,000 lb at an airfield 4,000 ft above sea level, in standard atmospheric conditions, both versions of the C-5 can land in only 4,000 ft, measured from a height datum of 50 ft on the final approach.

C-5A and C-5B Differences and Performance

Strengthened Wing. Even before the C-5A entered operational service in 1970, it had become clear—following the structural failure of a test specimen in mid-1969—that the mainplane would need to be significantly strengthened if the aircraft was to have a useful productive life and a realistic envelope of manoeuvre. After initial remedial measures, which offered no long-term solution, and following more detailed study of the problem, it was decided that the entire fleet of C-5As should be fitted with a much stronger wing torsion box. Accordingly, in a major and costly programme extending from 1980 to 1987, this component was entirely replaced; only the leading- and trailing-edge components (ie, slats, spoilers, flaps and ailerons) were retained. The new wing torsion box is manufactured mainly from a new aluminium alloy with much improved resistance to cracking and corrosion. Although it is nearly 19,000 lb heavier than the original, this penalty is more than outweighed (literally!) by the substantial increases in zero fuel weight, maximum payload and take-off weight which have been made possible by the strengthened wing. For example, as illustrated in Table 2.1, the maximum payload of the modified C-5A, in a flight envelope restricted to 2.25 g, is 245,000 lb—some 37,000 lb more than the original C-5A.

Differences. While retaining the same external and internal dimensions as the C-5A, the C-5B incorporates not only the former's strengthened wing but also a number

TABLE 2.1 *Summary of C-5 performance*

		Original C-5A (unmodified wing)	Current C-5A (new wing)	C-5B
		LB		
Basic weight		351,072	370,000	374,000
Maximum zero fuel weight	2.5 g	513,904	570,000	590,000
	2.25 g	558,904	615,000	635,000
Maximum payload	2.5 g	162,832	200,000	216,000
	2.25 g	207,832	245,000	261,000
Maximum take-off weight		769,000	837,000	837,000
Maximum in-flight weight	2.5 g	728,000	769,000	769,000
	2.25 g	769,000	840,000	840,000
Maximum landing weight	6 FPS	769,000	920,000	920,000
	9 FPS	635,850	635,850	635,850
		Nautical Miles (nm)		
Range with maximum payload	2.25 g	3,296	3,339	2,958

of other improvements, some involving modifications carried out on the C-5A during its period of service and others exploiting the availability of state-of-the-art avionics or new construction materials and techniques. For example, the C-5B's advanced avionics suite comprises a number of items already introduced or projected for the C-5A plus some newer equipment, including:

• MADAR II (an improved version of the malfunction detection system described earlier).
• Updated AFCS.
• Bendix AN/APS-133 digital weather/mapping radar with colour display.
• Three carousel IV(E) inertial navigation systems.
• Two AN/ARN-118 Tacan receivers.
• Two AN/ARN-127 VOR/ILS receivers.
• Two AN/ARC-186 VHF AM/FM radios.
• Two AN/ARC-190 HF SSB radios.
• Sundstrand Mk II ground proximity warning system.
• Crash-survivable flight data and voice recording system.

In construction terms, the C-5B has benefited from the availability of tougher aluminium alloys and other new materials, as well as from analysis of data collected during exhaustive testing of the new wing components. The net result is that the C-5B—despite having a higher basic weight—can carry 16,000 lb of payload more than the modified C-5A.

Summary of Leading Particulars

Flight deck crew: Two pilots and flight engineer.

External dimensions: Length   247.8 ft (75.53 m)
Wingspan   222.8 ft (67.91 m)
Height   65.1 ft (19.84 m).

Internal dimensions: Length (excluding ramp)   121.1 ft (36.91 m)
(cargo hold)   Length (including ramp)   144.6 ft (44.07 m)
                        Maximum width   19.0 ft (5.79 m)
                        Height   13.5 ft (4.11 m)
                        Volume (including ramps) 34,795 cu ft (985.3 m³).

Engines: All C-5 aircraft are now fitted with four General Electric TF39-1C turbofans, each producing 41,100 lb of thrust (43,000 lb with additional rating).

Performance: See Table 2.1.

Fuel capacity: 332,500 lb.

Cruise speeds: High speed-regime   0.79 Mach
                        Long-range regime   0.77 Mach

Service ceiling (615,000 lb gross weight): 35,750 ft.

C-141 Starlifter

Like the C-5 Galaxy, the C-141 Starlifter is manufactured by Lockheed and operated only by the USAF[2]. Although the Starlifter is much thinner (and some 80 ft shorter) than the wide-bodied Galaxy, the two types clearly share a common pedigree, each featuring the distinctive 'flying' T-tail, wing swept to 25° and four turbofan engines mounted on pylons beneath the wings.

The Starlifter has now been in operational service for over 20 years, having entered the USAF inventory in 1965 as the C-141A. MAC acquired 284 of these aircraft, effectively doubling its airlift capacity at a time when the demands of the war in Vietnam were putting renewed emphasis and pressure on strategic airlift. The advent of the Starlifter could therefore not have been more opportune, its combination of capacity and speed enabling it to reduce the cost of airlifting a ton of cargo from the east coast of the USA to Saigon by 50% and delivery time by one-third, compared with the best of the other strategic transports then in service.

PLATE 2.3.   C-141B Starlifter in flight.

Conversion of C-141A to C-141B

Despite the C-141A's impressive performance, it was increasingly found to suffer from one major weakness, namely an imbalance between the volume or cubic capacity of the cargo compartment and its load-carrying capability in terms of weight. This meant that, except where payloads were particularly dense—such as when pallets of ammunition were carried—the sheer bulk of many items filled up the hold well before the aircraft's maximum payload was reached, a phenomenon sometimes termed 'bulking out'. In an effort to redress this imbalance, and hence exploit more fully the Starlifter's weight-lifting potential, studies were set in hand which indicated that a 280-in (23.33-ft) stretch of the fuselage would produce an optimum relationship between the cost of the modification and a significant improvement in volume, whilst allowing retention of the existing wing, landing gear and engines. Recognising the long-term operational benefits and cost-effectivenes of such a modification, the USAF placed a contract with Lockheed in 1978 to stretch 270 Starlifters from 145 to 168.3 ft and to equip each of these aircraft with an aerial refuelling system to enable them to receive fuel from boom-equipped tankers. This programme, completed in 1982, added 240 sq ft of cargo floor space and 2,171 cu ft of volume, increasing the total capacity to 11,399 cu ft. The net effect was to increase the aircraft's volumetric cargo capacity by some 25%, enabling it to carry 13 instead of 10 standard 463L pallets. Put another way, the stretch programme provided MAC with the equivalent of 70 new aircraft without suffering any significant reduction in performance from the increase in basic weight, or needing any increase in the number of aircrews and maintenance personnel. The modified Starlifter was redesignated the C-141B.

General Description

The C-141B is an extremely versatile aircraft which can undertake a wide variety of strategic transport missions, including the airlift of passengers, cargo and a combination of both. Its major dimensions are illustrated in Figure 2.2.

Cargo Role. Designed for rapid and easy loading through the rear cargo doors, the passenger/cargo compartment is 10.3 ft wide, 9.1 ft high and 93.3 ft long. An additional 11.1 ft of loadable space is available on the aft ramp. Twelve standard pallets can be accommodated on the main floor and a further pallet on the ramp. Wheeled and tracked vehicles are loaded via the full-width ramp and auxiliary ground ramps (see Plate 2.4) and a winch is provided to assist the loading of both pallets and vehicles.

The aircraft has an integral roller conveyor and rail system to facilitate the handling of palletised freight. These rollers can be easily converted (in 10.5-ft sections) to flat floor areas for vehicles, thus providing flexibility in the loading of mixed vehicle/cargo configurations. Tie-down fittings of 10,000 and 25,000 lb capacity provide restraint for vehicles or unpalletised bulk cargo.

Trooping Role. The main cabin floor is designed to accommodate rearward facing passenger seats, canvas side-facing paratroop seats or stretchers (otherwise known as litters). Any of these configurations can be fitted in about an hour by a team of 12 personnel. A 'comfort' pallet, providing separate toilet and galley facilities for passengers, can also be readily installed in the forward section of the cabin. As indicated in Table 2.2, the number of personnel that can be carried varies according

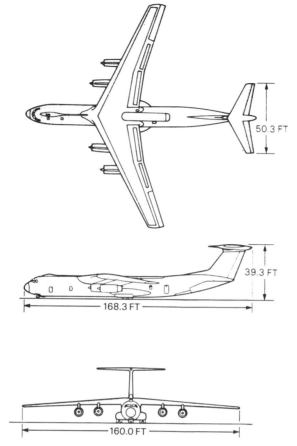

FIG 2.2.   Major dimensions of C-141B.

to the configuration and whether or not the comfort pallet is fitted.

Air-Drop Role. Another important feature of the Starlifter's versatility is its ability to air-drop both troops and equipment. Whether this signifies that the C-141B—or for that matter the C-5—can truly be considered a tactical as well as a strategic airlifter is a moot point. Arguably, the use of either of these labels should relate more to the *type* of task being undertaken than the *range* at which it is being performed, contrary to the conventional definitions of strategic and tactical operations. In any event, the distinction between strategic and tactical operations can easily become blurred especially when, as with the C-141B, tactical tasks form the final phase of a strategic mission. In view of its 'long legs', not to mention the availability of other aircraft such as the C-130 for the purely tactical role, it would make little sense to employ the C-141B on short- or medium-range air-drop missions.

Special equipment is needed for the air-drop role but this can be readily installed using quick-fit components. Equipment and supplies are usually despatched on platforms, which are 9 ft wide and vary in length from 8 to 24 ft in 4-ft increments. They can be dropped singly or in multiples, subject to a maximum total weight of

PLATE 2.4. Truck being unloaded from C-141B.

TABLE 2.2 *Passenger capacity of C-141B*

	With comfort pallet	Without comfort pallet
Troops	185	208
Paratroops*	149	168
Passenger	166	184

*Fewer paratroops can be carried because of their additional equipment.

74,000 lb (normal operations) or 94,500 lb (emergency operations). Individually air-dropped platforms must not exceed 35,000 lb. Equipment and supplies are despatched through the aft cargo doors, and paratroops through two doors positioned each side of the rear fuselage.

Landing Gear. There are ten wheels altogether. The main landing gear of eight wheels consists of two '4-wheel bogies', each mounting twin wheels forward and aft of the strut. The nose gear, which incorporates hydraulic steering, is a dual-wheel unit.

Airfield Operations. Although it cannot operate into short or unpaved airstrips, the C-141B is specifically designed to deliver its payloads of personnel and cargo into forward airfields which can be described as 'austere' by virtue of their inability to provide the range of support and servicing facilities usually found at more sophisticated airfields. Provided that the airfield's basic dimensions and load-bearing strength are adequate, and that there is handling equipment to deal with any freight carried on the roller conveyor system, the C-141B is virtually self-sufficient when

PLATE 2.5.   Starlifter dropping paratroops.

PLATE 2.6.   Starlifter dropping equipment platform.

operating away from main bases, thanks to the built-in maintainability and reliability of its systems and sub-systems. The minimal servicing required during the quick turn-rounds that are a feature of forward operations can easily be completed by the flight engineer (or crew chief, if carried) during off-loading or loading. Refuelling, if available and required, can usually be accomplished in under 30 min and the entire off-loading/loading process and turn-round in no more than an hour. The Starlifter is also its own best prime mover, with the ability to reverse if required. It is highly manoeuvrable on the ground, being able to execute a 180° turn in less than 140 ft of runway and turn onto 50-ft wide taxiways from 90-ft wide runways without requiring fillets.

Summary of Leading Particulars

Flight deck crew: Two pilots, navigator and flight engineer.
External dimensions: Length 168.3 ft (51.30 m)
Wingspan 160.0 ft (48.77 m)
Height 39.3 ft (11.98 m).
Internal dimensions: (cargo hold)
Length (including ramp) 104.4 ft (31.82 m)
Width 10.3 ft (3.14 m)
Height 9.1 ft (2.77 m)
Volume (including ramp) 9.190 cu ft (260.0 m³)
Engines: Four Pratt & Whitney TF-33-P-7 turbofans each providing 21,000 lb thrust.
Basic weight: 144,492 lb.
Maximum payload: 2.5 g–74,233 lb.
2.25 g–94,508 lb.
Fuel capacity: 153,352 lb.
Maximum take-off weight: 2.5 g – 323,100 lb.
2.25 g–343,000 lb.
Maximum landing weight: 10 fps – 257,500 lb.
6 fps – 343,000 lb.

Range with 94,508 lb payload: 2,500 nm.
Cruise speeds: High-speed regime 0.77 Mach
                      Long-range regime 0.74 Mach
Service ceiling (300,000 lb gross weight): 30,000 ft.

AN-124 Condor

As far as can be established from open sources, the Antonov AN-124 (code-named Condor by NATO) first flew in 1982 and began operations with Aeroflot, albeit on a very limited scale, in 1985. In early 1988, only some five aircraft were believed to have entered productive service, of which perhaps two had been delivered to the Soviet Air Force's *VTA*. The Condor seems to be intended mainly as a replacement for the *VTA*'s ageing fleet of AN-22s, although its initial appearance in Aeroflot livery suggests that it will also have commercial applications, probably including the movement of heavy and bulky freight between the industrial areas of the USSR and underdeveloped provinces in Siberia and the Soviet Far East. Civilian versions of the aircraft can also be expected to airlift outsize items of military hardware to distant overseas locations such as Angola, on the basis that a civilian carrier is likely to prove less provocative and pose fewer problems in obtaining overflight clearance than an overtly military flight.

When the AN-124 was first unveiled at the 1985 Paris Air Show, it was immediately hailed by the media as the world's largest aircraft, thus usurping the position hitherto occupied by the C-5 Galaxy. Some commentators also suggested, perhaps prematurely, that the AN-124 was not merely superior to the C-5 in size but also in design and performance. Such claims may well be true but need careful justification if the relative merits of these two strategic airlifters are to be objectively evaluated, especially in view of the customary Soviet reticence to provide the detailed information required

PLATE 2.7.   AN-124 in Aeroflot livery.

for any in-depth analysis of the AN-124's performance. Nevertheless, using the limited (and unsubstantiated) AN-124 data currently available, it is interesting to compare the two aircraft's respective dimensions and operating weights (see Table 2.3).

The data listed in Table 2.3 reveals that the C-5 is by no means as outclassed by the AN-124 as some of the initial publicity may have suggested. For example, while the AN-124 has a larger wingspan and higher fin than the C-5, the latter is 18 ft longer. As a result, the AN-124's cargo compartment is 3 ft shorter than the C-5's, a discrepancy which increases to 26.6 ft if the loading space on the C-5's rear ramp is included. (Cargo cannot be carried on the AN-124's rear ramp.) Even so, with a slightly higher and wider hold, the AN-124 has the edge in volumetric capacity and, thanks to other qualities (discussed later), a considerable payload advantage in terms of weight, being able to lift 69,750 lb more than the C-5B. Whether this additional capability can be fully exploited is another matter; as explained with regard to the C-141, airlifters frequently 'bulk out' before reaching their weight-lifting limit which can normally be utilised only when particularly dense loads are carried.

Design Features

Despite its generally greater dimensions, its heavier weight-lifting capability and its much larger fuel capacity, the Condor weighs only 11,000 lb more than the Galaxy. This impressive achievement is the result of skilful design based on the employment of new materials, technology and construction techniques.

Advanced Wing. One of the AN-124's most important features is its advanced wing, variously described in the West (where it originated) as 'supercritical' or 'aft-loaded'. Characterised by its flat upper surface and undercut trailing edge, this type of wing is longer than an orthodox wing of the same weight, thus providing both more lift and more space for fuel tanks. The wing also follows the Western trend away from complex high-lift devices. Instead, the AN-124 has full-span slats and slotted Fowler flaps in three sections on each side. In contrast with previous large Soviet aircraft, the wing has no vortex generators or fences.

Fly-By-Wire Controls. A valuable reduction in basic weight is also obtained by

TABLE 2.3 *Comparison of C-5 and AN-124 main dimensions and weights*

	C-5B Galaxy	AN-124 Condor
External dimensions		
Length	247.8 ft	229.7 ft
Height	65.1 ft	74.1 ft
Wingspan	222.8 ft	240.5 ft
Cargo Hold		
Length (excluding ramp)	121.1 ft	118.0 ft
Length (including ramp)	144.6 ft	
Maximum height	13.5 ft	14.4 ft
Maximum width	19.0 ft	21.0 ft
Fuel capacity	332,500 lb	485,000 lb
Operating weights		
Basic weight	374,000 lb	385,000 lb
Maximum payload	261,000 lb	330,750 lb
Maximum take-off weight	837,000 lb	893,000 lb
Range with maximum payload	2,958 nm	2,428 nm

PLATE 2.8.   AN-124 cargo compartment.

PLATE 2.9.   Rear ramp of AN-124.

use of an analogue fly-by-wire system which enables the aircraft to be flown with a high degree of accuracy and stability. Sometimes termed an 'artificial stability' system, this allows the designer to employ a relatively small, low-set, fixed-incidence tailplane, thereby further reducing drag, structural weight and system complexity.

Use of New Materials. Weight has also been saved by the extensive use of new materials, including glass-fibre reinforced plastic (GRP) and carbon-fibre composites (CFC). Such material is not, of course, employed in the primary airframe structure or control surfaces, but is incorporated as widely as possible elsewhere. For example, the fairing on the main landing gear is made from GRP and the doors of the nose and main undercarriage, as well as the large cargo doors aft, are manufactured from CFC. Altogether, it is estimated that the GRP and CFC components are some 4,000 lb lighter than if they had been constructed in metal. The designers have saved further weight by building the cargo floor from titanium alloy, a very hard-wearing but extremely expensive material. As the world's largest producer of titanium, the USSR can presumably afford to use it on a scale and for a purpose which would be too costly to contemplate in the West.

Landing Gear. The landing gear has 12 pairs of wheels to spread the aircraft's enormous weight: two pairs under the nose and a further five pairs on each side of the fuselage. It seems likely that the flotation (ie, suspension) of the gear is sufficient, like that of the C-5, to permit operations from semi-prepared (though probably not soft) surfaces. On the ground, a kneeling mechanism allows the height of each main unit to be adjusted in order to tilt the aircraft for loading or unloading through either nose or tail. The AN-124's visor nose is much like that of the C-5 except that, when it is opened, it exposes the nosewheels which then retract in the usual way, allowing the fuselage to settle onto a pair of retractable 'feet'. A three-piece folding ramp is then extended at a shallow angle for vehicle loading. The visor-opening and ramp-unfolding sequence takes about 7 min.

Double-Decker Configuration. Like the C-5, the AN-124 has an upper and lower deck.

- Cargo compartment:
  As mentioned, the floor of the cargo compartment is constructed from titanium alloy. There is no roller conveyor system as in the C-5; instead, loads are manoeuvred within the aircraft by four overhead cranes, mounted in pairs on twin rails running the length of the cargo hold ceiling (see Plate 2.11). Each crane has two hoists, each of which is capable of lifting five tonnes and each of which has lateral freedom of movement. This system is typical of Soviet design and seems to offer significant advantages in terms of speed and ease of handling.
- Flight deck and upper cabin:
  The upper level is divided at the wing into two main sections, the rear area providing seats for 88 passengers. The forward sections consists of the flight deck with six crew stations, galley, toilet and a rest area for a further crew of six. Each upper section is reached by a retractable ladder from the cargo hold. The flight deck crew consists of two pilots, a navigator, two flight engineers and a radio operator. This is a large crew by contemporary Western standards but overmanning seems to be as endemic in aviation as in many other walks of Soviet life. Even so, the employment of two flight engineers is most unusual. Presumably

PLATE 2.10.   AN-124 with nose visor raised.

PLATE 2.11.   Overhead cranes to assist freight-handling in AN-124.

they are considered necessary to handle the aircraft's relatively complex systems, and to assist with turn-rounds when away from base.

Flight Instruments and Avionics. The pilot's instrument panels are conventional and fairly austere. For example, their engine instruments show only the overall pressure ratio and throttle position, the remaining gauges being positioned on a more comprehensive engine panel at one of the flight engineer stations. Unlike many aircraft now produced in the West, the AN-124 does not have an electronic flight instrumentation system (EFIS) but relies mainly on orthodox electromechanical equipment. However, it does have a digital computer, similar to the C-5's MADAR, which continuously monitors aircraft systems and detects and diagnoses failures either in flight or on the ground. The cathode ray tube (CRT) display and controls for this system are located between the flight engineers' consoles. The Condor's navigation fit includes a triple inertial navigation system (INS), Loran, OMEGA and a weather/mapping radar, as well as a moving map display.

Engines. Having been unsuccessful in various attempts to acquire Western turbofans, the USSR was finally obliged to develop its own powerplant for the AN-124. As far as can be judged, the four Lotarev D-18T engines, each providing 51,650 lb of thrust and fitted with highly effective reversers, seem well able to perform their onerous task. Similar to the Rolls-Royce RB-211 in terms of its number of stages per spool and limited use of variable geometry, the D-18T's bypass ratio is rather higher in order to harness its performance to the aircraft's optimum cruising speed of 0.75 Mach. One of the D-18T's most notable features is its design life of 18,000 hours

PLATE 2.12. AN-124 flight deck.

PLATE 2.13.   AN-124 pilots' instrument panel.

with intervals of 4,000–6,000 hours between major servicing. This suggests that Soviet aero-engine designers may have finally overcome the problem of durability and reliability which, in the case of previous engines, dictated far more frequent intervals between major overhauls. It is also worth noting that the Condor's engines produce a total thrust of 206,600 lb—some 25% more than the Galaxy's total thrust of 164,000 lb. This is another important factor in the Condor's higher weight-lifting capacity.

Overall Assessment of AN-124

The AN-124 is still relatively new and information concerning the role and performance of the handful of aircraft currently in service remains fairly sparse. Meanwhile, it will inevitably continue to be compared with the Galaxy. While the Galaxy is slightly smaller overall and can carry less payload and fuel, it is still an impressive airlifter by any standards. Moreover, despite its older design, it has several important advantages. For example, it can air-drop cargo (which the AN-124, lacking a roller conveyor system, cannot); and unlike its Soviet rival, the C-5 can extend its range indefinitely by its ability to refuel in the air. This factor should not be overlooked when comparing ranges and payloads. Nevertheless it seems reasonable to conclude that the AN-124 is a highly effective airlifter with the ability to operate in the outsize/heavy cargo role from a large number of airfields. Although it has been seen only in the West in Aeroflot livery to date, the AN-124's military potential is obvious and will further strengthen the USSR's already substantial capacity to mount strategic transport operations.

Summary of Leading Particulars

Flight deck crew: Two pilots, navigator, two flight engineers and radio operator.
Dimensions: See Table 2.3.
Engines: Four Lotarev D-18T turbofans, each providing 51,650 lb thrust.
Performance: See Table 2.3.
Typical cruise speed: 0.76 Mach.
Cruise altitude: 30,000–39,000 ft.

AN-22 Cock

Now well over 20 years old, the Antonov AN-22 turboprop freighter (code-named Cock by NATO) created a sensation when it first appeared in the West at the Paris Air Show in 1965. At that time, it was the world's largest and heaviest aircraft, a position which it retained for three years until the first flight of the C-5 Galaxy in 1968. About 100 AN-22s were eventually produced with the Soviet Air Force and Aeroflot each receiving some 50 aircraft. All were built to the same basic specification, thus enabling the civil variants to be employed on military tasks when so required. Such flexibility and interchangeability—made possible by the USSR's system of centralised planning and resource management—is a traditional feature of the Soviet concept of airlift operations. The advantages are manifold; for example, not only can civilian aircraft be rapidly diverted onto military tasks when unexpected contingencies arise but they can also be used for routine military missions. This latter option is especially valuable in some overseas scenarios where the use of commercially operated aircraft may be politically preferable to military airlifters. Aeroflot's fleet of AN-22s has certainly been extensively used for military or quasi-military tasks, including the airlift of arms and equipment to client states in Africa, the Middle East and Central America.

There is a strong family resemblance between the strategic AN-22 and its elder cousin, the tactical AN-12 (see Chapter 4). To some extent, the AN-22 is a scaled-up derivative of the AN-12, retaining the latter's anhedral outer wings, distinctive glazed nose and ability to operate from short, austere airfields. There are some significant differences, however, notably in the provision of contra-rotating propellers and in the design of the rear fuselage and tail. As noted earlier, designers of large military airlifters have relatively little room for innovation if such aircraft are to do the job for which they are intended. The typical arrangement of aft ramp and doors with upswept tail poses particular structural problems in so far as this configuration tends to generate considerable drag and aerodynamic loading on the rear fuselage, especially when the aircraft in question is as large as the AN-22. The Antonov

PLATE 2.14.   AN-22 Cock taxying.

bureau's solution was to fit a twin-fin tail topped off with anti-flutter devices, instead of the single fin used on the smaller AN-12.

Short Field Performance

For its size—it is shorter in length and wingspan than the AN-124 by only some 40 and 29 ft respectively—the AN-22 has an impressive take-off and landing performance. For example, at its maximum take-off weight of 550,000 lb, it requires a ground roll of only 5,000 ft at sea level in ISA conditions. This short-field capability is achieved mainly by the use of 'blown' flaps, the large contra-rotating propellers (20 ft 4 in diameter) washing a powerful slipstream over the inner wings, thereby greatly enhancing lift at low airspeeds. In order to exploit this capability in operations from austere or unprepared airfields, the AN-22 is equipped with an extremely robust undercarriage, the port and starboard main gear each consisting of three double-wheel units with very large low-pressure tyres. These main gear units are housed in massive 'blisters' protruding from each side of the fuselage.

Other Design Features

The AN-22 is designed and used primarily as a freighter although it also has a small cabin aft of the flight deck with seats for about 30 passengers. Behind this cabin, from which it is separated by a bulkhead and connecting doors, is the spacious cargo hold (see dimensions below). With a reinforced titanium floor, tie-down fittings and full-width rear loading ramp, the main hold can accommodate most items of Soviet Army equipment including heavy engineering plant and T-62 main battle tanks. Rails running along the entire ceiling of the cargo hold carry mobile gantries equipped with electrically operated hoists which, together with integral winches, ease the handling of heavy containers and other items of freight. When the ramp is lowered, a large door—which otherwise forms the underside of the rear fuselage—retracts upward to permit the loading of high vehicles. Air-dropping can also be conducted through this rear door although, in practice, the aircraft is seldom used in this role.

PLATE 2.15.   AN-22 Cock in flight.

Overall—despite its age, obsolescent systems and relatively slow cruising speed—the AN-22 is still an effective airlifter by virtue of its ability to carry heavy and bulky loads over substantial distances, operating if necessary from short, unpaved airfields.

Summary of Leading Particulars

Flight deck crew: Two pilots, navigator, flight engineer and radio operator. (The navigator's station is in the nose beneath the flight deck.)

External dimensions:	Length	190.0 ft (57.92 m)
	Wingspan	211.3 ft (64.40 m)
	Height	41.2 ft (12.54 m).
Cargo hold dimensions:	Length	108.3 ft (33.0 m)
	Maximum width	14.4 ft (4.4 m)
	Maximum height	14.4 ft (4.4 m).

Engines: Four Kuznetsov NK-12MA turboprops, each driving a pair of 4-blade contra-rotating propellers.
Basic weight: 251,325 lb (114,000 kg).
Maximum payload: 176,350 lb (80,000 kg).
Maximum fuel: 94,800 lb (43,000 kg).
Maximum take-off weight: 551,160 lb (250,000 kg).
Range with maximum fuel and 100,000 lb payload: 5,900 nm.
Range with maximum payload: 2,692 nm.
Cruise speed: 278 kt.
Service ceiling: 25,000 ft

IL-76 Candid

Designed in the late 1960s to meet a joint military/civil requirement, the Ilyushin IL-76 (code-named Candid by NATO) first flew in 1971 and entered service some four years later. The first Soviet airlifter to have underslung engines, the IL-76 was conceived not only as a strategic airlifter but also as the long-term replacement for the USSR's ageing fleet of AN-12s. As can be seen from Plate 2.16, it is similar in appearance to the C-141B Starlifter, with which it is often compared. However, while the IL-76 performs much the same spectrum of missions as the C-141B, its short-field performance is superior to that of its American rival.

Like the AN-12 and AN-22, the IL-76 is in service with both the *VTA* and Aeroflot. With production still continuing, about 300 aircraft have been delivered to *VTA* squadrons to date, whilst a further 50 are operated by Aeroflot; the latter are effectively an immediately available military reserve. Outside the USSR, the IL-76 is in service with the air forces of Czechoslovakia, Poland, India and Iraq. In common with the AN-12 and AN-22, the IL-76 was specifically designed at the outset for operations from short and semi-prepared fields. To this end, the aircraft is equipped with a variety of high-lift devices including ten-segment slats over virtually the entire wing leading edge; two and three segments on each inner and outer panel respectively; and triple-slotted flaps over approximately three-quarters of each wing between the root and inboard edge of the aileron. There are also 16 upper surface spoilers forward of the flaps (four on each inner and outer section of the wings) to enhance handling and roll-control at low speeds.

Landing Gear

The IL-76 is equipped with a heavy-duty landing gear, consisting of a nose unit and two main units in tandem on each side of the fuselage, all units having four large wheels in line abreast on single axles. After the gear is lowered, all undercarriage doors are automatically closed to prevent the ingress of mud, snow or ice into the wheel wells; after take-off, the main units retract into two large ventral fairings. The crew can adjust tyre pressures in flight (from 2.5 to 5 bars) to suit different changing landing conditions. This is an extremely useful feature which, coupled with the rugged landing gear, enables the aircraft to operate from a variety of rough or semi-prepared surfaces.

PLATE 2.16.   IL-76 Candid in flight.

Cargo Hold

The IL-76 has an excellent cargo hold, both in terms of volume (see dimensions below) and role equipment. Constructed of titanium alloy, the reinforced floor is fitted with folding rollers, with winches, overhead cranes and hoists also provided in typical Soviet fashion to facilitate the handling of containerised freight and other heavy items. The configuration of the hold can be readily adapted to permit the carriage of various cargo/passenger permutations by fitting up to three modules, each able to accommodate passengers, stretcher patients (with aeromedical attendants) or cargo as required. Wheeled and tracked vehicles can self-load expeditiously via the wide rear ramp.

Performance

With similar overall dimensions to those of the C-141B, and with a virtually identical wing area, the IL-76 can carry rather more payload than the US aircraft over comparable distances. Moreover, while performance data is difficult to obtain and corroborate, it is believed that the IL-76 is capable of heavy all-up weight operations from 6,000-ft airfields (at sea level in standard atmospheric conditions). Like its American counterpart, the aircraft is also an effective air-dropping platform for both paratroops and heavy equipment. Its extensive employment in this role may explain why, like the AN-12, the IL-76 has a tail-gunner's station, though the rationale for armament in any transport aircraft is difficult to understand. In the case of the IL-76, this feature might lend itself to conversion as a flight refuelling operator's station if, as seems likely, the aircraft is modified for the tanker/transport role. Meanwhile, the standard transport version[3] is not equipped even to receive fuel in flight, this being its major disadvantage *vis-à-vis* the Starlifter. In all other respects, however, the IL-76 is by any standards an impressive airlifter which seems set to remain the backbone of the *VTA*'s strategic—and tactical—transport fleet well into the next century.

PLATE 2.17   IL-76 Candid on ground.

Summary of Leading Particulars

Flight deck crew: Two pilots, navigator, flight engineer and radio operator. (The navigator's station is in the glazed nose compartment below the flight deck.)

External dimensions: Length   152.8 ft (46.59 m)
                             Wingspan   165.7 ft (50.50 m)
                             Height   48.4 ft (14.76 m).

Cargo hold dimensions: Length (excluding ramp)   65.6 ft (20.00 m)
                             Length (including ramp)   80.3 ft (24.50 m)
                             Maximum width   11.2 ft (3.40 m)
                             Maximum height   11.3 ft (3.46 m).

Engines: Four Soloviev D-30KP turbofans.
Basic weight: 135,000 lb (62,000 kg).
Maximum payload: 105,820 lb (48,000 kg).
Maximum fuel: 131,000 lb (59,420 kg).
Maximum take-off weight: 418,875 lb (190,000 kg).
Range with 88,000 lb payload: 2,700 nm.
Maximum range (with normal reserves): 3,600 nm.
Cruise speed: 0.75 Mach.
Normal cruising altitudes: 30,000–39,000 ft.
Service ceiling: 42,000 ft.
Armament (when fitted): Twin radar-directed 23 mm cannon in manned tail turret.

OTHER STRATEGIC AIRLIFTERS

As indicated earlier, only the USA and USSR can afford to operate large fleets of strategic transports, the more important of which have been surveyed in the preceding sections. Other states with a more modest strategic capability include the UK with its Tristar tanker/transports (covered in Chapter 3) and VC-10s; and Canada, Israel and the Federal Republic of Germany, all with various versions of the B-707. The characteristics and capabilities of the VC-10 and B-707 are briefly examined below.

VC-10 CMk 1

The RAF operates one squadron of 13 VC10 CMk 1 aircraft. Acquired new in the mid-1960s as military variants of the civilian VC-10 airliner, they have performed well over the years, playing an important role in several long-range operations, notably the airlift of personnel and supplies between the UK and Ascension during the Falklands war of 1982. However, the VC-10 is based on a design conceived in the 1950s, and its airlift potential and capabilities compare unfavourably with custom-built military transports such as the C-141B Starlifter. Moreover, the VC-10 is very noisy in environmental terms and its engines are much less fuel-efficient than modern turbofans. Nevertheless, the aircraft is expected to continue in service until at least the end of the century.

Equipped with a strengthened floor and a cargo door on the port side of the foward fuselage, the VC-10 CMk 1 can be used to carry freight, passengers or various combinations of both. In the full passenger configuration, up to 150 troops can be

PLATE 2.18.   VC-10 CMk 1 in flight.

carried while, in the aeromedical role, the aircraft can accommodate up to 78 stretcher patients or 61 sitting patients plus nine on stretchers. The VC-10 CMk 1's freight-carrying capacity is limited not only by the maximum payload restriction (46,000 lb) but also by the relatively small dimensions of its cargo door and main cabin (see below). However, the aircraft can carry its maximum payload for 3,380 nm, a range which compares well with those of more modern airlifters and which can be further extended by in-flight refuelling. Using this technique, and carrying a mixed payload of personnel and freight, a VC-10 CMk 1 made the first direct flight from the UK to the Falkland Islands on 19 December 1987. Refuelling twice *en route*, the aircraft covered the distance of 7,200 nm in 15 hr 45 min.

Summary of Leading Particulars

Flight deck crew: Two pilots, navigator and flight engineer.

External dimensions:	Length	158.7 ft (48.36 m)
	Wingspan	146.2 ft (44.55 m)
	Height	39.5 ft (12.04 m).
Cargo door:	Height	7.0 ft (2.13 m)
	Width	11.7 ft (3.55 m).
Main cabin:	Length	92.3 ft (28.14 m)
	Maximum width	11.5 ft (3.50 m)
	Maximum height	7.4 ft (2.26 m)
	Volume	6,700 cu ft (190 m^3).

Engines: Four Rolls-Royce Conway Mk 250 turbofans, mounted in lateral pairs on each side of rear fuselage. Each engine produces 21,800 lb static thrust. Only the outboard engines have reversers.

Basic weight: 149,868 lb (67,980 kg).

Maximum payload: 46,000 lb (20,865 kg).

Maximum fuel: 131,000 lb (59,420 kg).

Maximum take-off weight: 323,000 lb (146,510 kg).
Range with maximum payload: 3,380 nm.
Range with maximum fuel: 5,000 nm.
Cruise speed: 0.81 Mach.
Cruise altitude: 30,000–39,000 ft.
Service ceiling: 42,000 ft.

Boeing 707

Although the prototype first flew as long ago as 1954, the Boeing 707—many derivatives of which are still in both commercial and military service—has a more respectable airlift performance that its date of origin might suggest. An improved version, the B-707-320C Convertible, specifically designed for passenger/cargo or all-cargo operations, entered airline service in 1963. Five of these aircraft, with the military designation CC-137, were subsequently delivered to the Canadian Armed Forces in 1970/71, primarily for the strategic airlift role but with two being equipped for a secondary tanker mission using wingtip hose and drogue pods. The description and performance outlined below refer to the CC-137.

Although its basic weight and dimensions are very similar to those of the VC-10 CMk 1, the CC-137 can carry almost twice as much payload at identical speeds over virtually the same distance. The CC-137 can also carry up to 219 passengers compared with the VC-10's maximum of 150. As with the VC-10 CMk 1, however, the CC-137's freight capacity is limited both by its narrow-body fuselage and by the size of the side-loading cargo door (see Plate 2.20 and summary of particulars below). A cargo-handling system is fitted to a section of the strengthened floor of the upper compartment which, with a total capacity of 5,693 cu ft, can accommodate up to thirteen 10 ft 5 in × 7 ft 4 in pallets. The lower hold, with a volume of 1,700 cu ft, can be used for unpalletised cargo and/or passenger luggage. Like the VC-10 CMk 1, the CC-137 is relatively noisy and unable to operate from short or unprepared airfields.

PLATE 2.19.   CAF CC-137.

PLATE 2.20.   Loading in progress through cargo door of CAF CC-137.

Summary of Leading Particulars

Flight deck crew: Two pilots, navigator and flight engineer.

External dimensions:	Length	152.9 ft (46.61 m)
	Wingspan	145.8 ft (44.42 m)
	Height	42.4 ft (12.93 m).
Cargo door:	Height	7.7 ft (2.34 m)
	Width	11.2 ft (3.40 m)
	Height of sill	10.5 ft (3.20 m).
Main cabin:	Length	111.3 ft (33.93 m)
	Maximum width	11.7 ft (3.55 m)
	Maximum height	7.7 ft (2.34 m).

Engines: Four Pratt & Whitney J T3D-7 turbofans, each providing 19,000 lb static thrust.

Basic weight:	Passenger role	146,400 lb (66,406 kg)
	Cargo role	141,100 lb (64,000 kg)

Maximum payload: 88,900 lb (40,324 kg).
Maximum fuel: 145,200 lb (65,860 kg).
Maximum take-off weight: 333,600 lb (151,315 kg).
Range with 80,000 lb payload: 3,150 nm.
Range with maximum fuel: 5,000 nm.
Cruise speed: 0.81 Mach.
Cruise altitude: 30,000–37,000 ft.
Service ceiling: 39,000 ft.

Questions

1. List five basic requirements of a strategic airlifter.

2. (a) What is the primary mission of the C-5 Galaxy?

   (b) What method of in-flight refuelling does the C-5 employ?

   (c) What is the function of the C-5's MADAR system?

3. Both the C-5 and the AN-124 have front and rear access to the cargo hold. What advantages does this confer?

4. The original version of the Starlifter (the C-141A) was prone to 'bulk out'. Explain what is meant by this term, and indicate how the problem was alleviated.

5. Under what circumstances is the C-141B likely to be employed on tactical missions?

6. What are the main advantages and disadvantages of the AN-124 compared with the C-5B?

7. Some types of strategic airlifter are used by both Aeroflot and the *VTA*. What benefits does the USSR derive from this policy?

8. (a) What are the aerodynamic drawbacks of the 'upswept tail/rear ramp and doors' configuration usually found in a military airlifters, and how is the problem overcome in the AN-22?

   (b) For such a large aircraft, the AN-22 has impressive take-off and landing performance. How is this achieved?

3

Tanker/Transport Operations

DEVELOPMENT OF AIR-TO-AIR REFUELLING

Although under more or less continuous development since the 1930s, the concept of air-to-air refuelling (AAR) is a relatively recent feature of military air operations. Originally conceived as little more than a means of extending the range and/or payload of fighters and bombers, AAR remained relatively unrecognised until the mid-1950s when, driven by the politico-military imperatives of the post-war era, the USA embarked on a major programme to produce the world's first fleet of dedicated turbofan-powered tankers. These were the highly acclaimed KC-135 Stratotankers of which 732 were delivered to the USAF in the ten-year period following the aircraft's first flight in 1956.

Although a derivative of—and similar in size and appearance to—the Boeing 707 airliner, the KC-135 is in fact quite a different aircraft, having been designed to military specifications and constructed to operate at higher gross weights than its commercial cousin. The fuel tankage is located in the wings and beneath the fuselage floor, leaving space in the main cabin for up to 80 passengers. However, since the Stratotanker cannot carry any significant payload without seriously impairing its primary value as an airborne filling station, it is not a tanker/transport in the strict sense of that term, use of which implies at least some ability to undertake both roles simultaneously. Hence the KC-135—some 600 of which are still in operational service today—remains essentially a dedicated tanker, with its modest transport capability available only for emergency use. The RAF's squadron of VC-10 tankers, although converted from former civilian airliners, is also employed exclusively in the tanker role, retaining virtually no transport capability.

Neither the KC-135 nor the VC-10 tanker will be considered further in this chapter, the remainder of which will focus on the systems and techniques of those aircraft that *are* able to complete both tanker and transport tasks on the same mission.

WIDER APPLICATIONS OF AAR

The advent of the highly effective KC-135 did much to stimulate a greater awareness of the operational value of AAR, which in recent years has been seen to confer a far wider degree of both strategic and tactical flexibility than was at first envisaged. In particular, air commanders around the world have increasingly come to recognise that, far from being relevant only to air defence and strike/attack scenarios, AAR applications can now be exploited to cover the full spectrum of operational tasks, especially in the maritime, airborne early warning and transport roles. Several events in recent years have helped to reinforce this growing perception of AAR's broader potential. For example, during the Falklands war of 1982, AAR proved absolutely crucial to the RAF's ability to mount missions into the operational area from its nearest base at Ascension Island (see Figure 1.2). Among the many AAR-supported missions flown from Ascension Island during this conflict were the historic attacks by Vulcan bombers on the Argentine-held airfield at Stanley on East Falkland. Although these attacks inflicted little physical damage, their real significance lay in their strategic implications for, and psychological impact upon, the enemy. In particular, these raids demonstrated that, by the adroit exploitation of AAR, the RAF could reach out over immense distances and deliver further attacks if so ordered. As a result, the Argentinians were forced to divert assets to the defence of their mainland air bases (which they now realised were also vulnerable) which might otherwise have been deployed in an offensive role against the British Task Force.

As well as supporting these bombing raids, AAR forces were also used to sustain a number of other air operations in the South Atlantic, including some involving airlift tasks. During hostilities and for some time afterwards, probe-equipped C-130s (supported initially by RAF Victor tankers and later by other C-130s rapidly converted to the tanker role) were the UK's only transport aircraft capable of reaching the Falkland Islands with sufficient fuel to divert or recover safely to Ascension. Subsequently, it was AAR that enabled the RAF to operate a regular and conspicuously successful 'airbridge'—again involving converted C-130s—between Ascension Island and the Falklands, an operation which for nearly three years continued to provide the only air link until the completion of a new airport at Mount Pleasant in 1985 opened up the route for regular Tristar flights.

An equally impressive demonstration of AAR's potential for extending the reach—

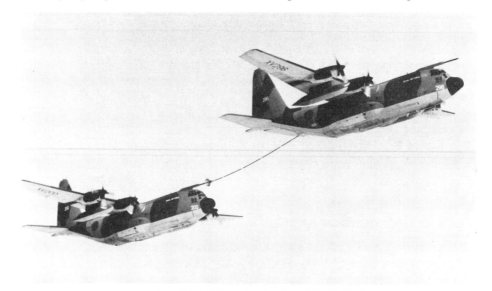

PLATE 3.1.   RAF C-130 refuelling from RAF C-130K.

PLATE 3.2.   Nimrod maritime patrol aircraft refuelling from RAF C-130K.

while reducing the timescale—of airlift operations was provided during Exercise REFORGER in 1982 when 20 C-141 Starlifters completed non-stop return flights from Pope and Charleston air force bases in the south-eastern USA to dropping zones in southern Germany, a round trip of some 10,000 miles. (See Figure 3.1).

Refuelling outbound and inbound over Canada from seven KC-10 Extenders, and over the UK from 20 KC-135 Stratotankers, the C-141s dropped 1,000 paratroops and several hundred tons of equipment directly into the exercise area. The following year, a force of Starlifters again exploited AAR to accomplish an even longer non-stop air-drop mission, this time involving a round trip of over 12,000 miles between the USA and Egypt during Exercise BRIGHT STAR 83.

DEVELOPMENT OF AAR AS AN AIRLIFT 'FORCE EXTENDER'

Collectively, these US exercises and the UK's operations in the South Atlantic confirmed once and for all that AAR was not merely a means of improving the range and endurance of air defence and strike/attack aircraft, but could also be particularly effective in enhancing the reach, scope and flexibility of airlift operations. Moreover, the operational and economic benefits of using dual-capability tanker/transport aircraft to support long-range deployments of fast jet combat units have become increasingly evident, since this has allowed the latter's groundcrew and support

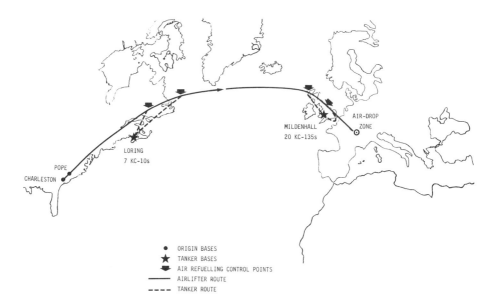

FIG 3.1.   Diagram of Exercise REFORGER 1982.

equipment to be carried in the same aircraft that are used to dispense fuel *en route*. In short, the growing use of in-flight refuelling by airlifters and the parallel development of tanker/transports have served jointly to underline the relevance of AAR to transport operations. The net effect has been to demonstrate that the concept of AAR as a 'force extender' or 'force multiplier' has far wider applications and implications than originally anticipated. In practical terms, this means that airlift potential can be enhanced by AAR to the substantial benefit of all other air and surface forces deployed, supported or redeployed by air transport operations.

AAR SYSTEMS AND TECHNIQUES

Before considering the performance of tanker/transports in more detail, it is necessary to examine the AAR systems which are available to such aircraft. Basically, there are only two in-flight refuelling systems, the flying boom method and the probe and drogue method. Developed by the Boeing Aircraft Company, the flying boom system is used only by the USAF[1], with the USN, RAF and other AAR operators around the world preferring the probe and drogue method pioneered by the UK company Flight Refuelling Limited.

Flying Boom System

After initially experimenting with the probe and drogue system, the Boeing Company opted to develop the flying boom system which, in their view, offered a superior rate of fuel transfer while placing less demand on the pilot of the receiving aircraft. Under this system, illustrated in Plate 3.3, a specialist operator lowers and extends a telescopic boom from the rear fuselage of the tanker. Then, from his special control station in the tail section, the operator uses small but sensitive aerofoil surfaces to 'fly' the male nozzle at the end of the boom into the female socket on the receiver aircraft, usually situated just aft of the receiver pilot's cockpit. When the nozzle is securely engaged in the socket, a valve automatically opens to allow fuel to flow. One of the advantages claimed for this system is that the receiver pilot has only to formate on the tanker within certain prearranged parameters—in theory, a relatively straightforward task though in practice a manoeuvre requiring no mean level of skill, especially when the receiver is a large aircraft—while the boom operator effects the actual link-up and controls the transfer process.

PLATE 3.3. USAF KC-10 transferring fuel to SR-71 by flying boom.

Probe and Drogue System

Instead of a telescopic boom, the probe and drogue system employs a flexible hose trailed from the tanker, and a probe instead of a socket on the receiver aircraft.

The probe, which may be fixed or retractable, is fitted with a male nozzle mounted in such a position as to enable the pilot to guide it accurately into position. The hose is carried on a rotatable drum unit mounted either inside the tanker's fuselage or in an underwing pod. The inboard end of the hose is connected to the tanks from which fuel is dispensed, whilst the other end culminates in a cone-shaped drogue or 'basket' incorporating a female reception coupling device. When the tanker trails the hose, the drogue acts not only as a stabiliser in the turbulent airflow but also as an aiming point for the receiver's probe. At normal refuelling speeds, the drag of the extended hose and drogue is such that, if permitted to run out unchecked, damage to both hose and drum would occur. For this reason, the hose drum unit's driving motor is used to restrain the hose while it is being deployed as well as to wind it in after refuelling has been completed.

To make contact, the receiver pilot formates upon the tanker and aims his probe at the trailing drogue. When the probe makes positive contact with the reception coupling at the apex of the drogue, cut-off valves open on each side of the connection and the nozzle is gripped by rollers mounted on spring-loaded toggle arms which lock all elements into position, thereby preparing the way for the transfer of fuel once the tanker's main cock is opened. On tanker and tanker/transport aircraft, the fuel transfer system and hose drum unit are controlled from a special panel on the flight deck, usually at the flight engineer's station. In order that the flight engineer can visually monitor the AAR sequence, an optical system is provided using either a simple periscope or—as in the case of the Lockheed Tristar tanker—closed circuit television. Mounted under the rear fuselage just forward of the drogue stowage tunnel, the Tristar's television camera provides a field of view extending to 345° horizontally and 33° vertically. This excellent coverage not only affords a picture of what is happening immediately to the rear of the aircraft during AAR, but also

PLATE 3.4.   RAF Tristar with hose and drogue deployed.

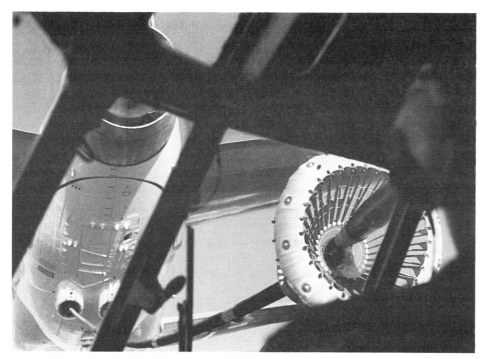

PLATE 3.5.   RAF C-130 about to receive fuel from a Tristar. The probe has not yet locked into the drogue.

includes the wing-pod dispensers and allows the crew to inspect the landing gear should this be necessary.

The impact of the probe as it engages the drogue or basket opposes the drag forces on the hose and, as the hose drum drive mechanism continues to exert a positive torque in the direction of wind-in, a few metres of hose are usually wound in automatically. The receiver pilot then adjusts his throttles to synchronise his speed with that of the tanker. Any discrepancy in speed between the two aircraft during refuelling will cause the hose to wind in or out, thus maintaining a constant hose tension without excessive sag or strain. Nevertheless, the nozzle has a 'weak link' section which, in the unlikely event of abnormal stresses, is designed to fracture, leading to rapid disengagement. Should this happen, the broken-off section of the probe would be retained in the reception coupling and hence would not endanger the receiver aircraft. It is also important to note that the toggle arms and nozzle shell are so designed that the load required to withdraw the probe is substantially greater than the entry load. This prevents premature breakaway during refuelling, yet does not restrict deliberate disengagement when the receiver aircraft reduces speed to the point at which the required pull-free load is exerted. When the probe is disengaged, all the valves close immediately to preclude any wastage of fuel.

Comparison of Systems

Each of the AAR systems described above has its merits—and its critics. The chief advantage of the flying boom method is its higher flow-rate (ie, the rate at which fuel can be transferred from tanker to receiver). While flow-rate is less critical to airlift or tanker/transport operations, it is an especially important consideration in scenarios involving combat aircraft, where the higher the rate of transfer, the quicker the AAR phase can be completed, thereby minimising the vulnerability of both tanker and receiver.

On the other hand, only one receiver aircraft at a time can be flight-refuelled by a tanker equipped solely with the flying boom system, a restriction which to some extent cancels out the high flow-rate. Probe and drogue tankers do not suffer from this limitation since they can be equipped to dispense fuel to three receivers simultaneously, by trailing one hose from the fuselage and further hoses from each of two wing pods. Although dual-system capability has been provided in the USAF's KC-10 Extenders by fitting a single hose and drogue system adjacent to the boom installation, this still does not allow more than one receiver to be flight-refuelled at any one time.

Moreover, aircraft which operate the flying boom system require an additional crew-member to act as boom operator. Whether this is an advantage or drawback depends on one's point of view. Although it can be fairly claimed that the operator's excellent vantage point at the rear of the tanker, and personal control of the entire transfer process, constitute a positive safety factor, some AAR experts consider that the rigid construction of the boom is inherently more hazardous to the receiver than the flexible hose of the probe and drogue system. Flight safety considerations have also generated some debate concerning the respective demands placed upon receiver pilots by each system. As already explained, a pilot receiving fuel via a flying boom has 'merely' to formate upon the tanker, leaving the boom operator to complete the link-up. However, once the transfer process is under way, the receiver pilot must maintain a very stable position relative to the tanker, in contrast with the pilot of a probe-equipped aircraft who, although responsible for initial as well as subsequent engagement, at least enjoys some modicum of manoeuvre in the vertical, horizontal and lateral axes.

Overall, it is difficult to state categorically that one system is superior to the other. Although used only by the USAF, the flying boom method is more widely employed in absolute terms simply because the USAF's AAR resources are substantially larger than those of all other nations put together. Nonetheless, some weight must be attached to the fact that every other air arm throughout the world, including that of the USN, has opted for the probe and drogue system, presumably because it is considered to offer both operational and economic advantages. The provision of underwing hose and drogue pods on its newly acquired KC-10 Extenders would appear to indicate that this point has not been lost on the USAF which now seems to have recognised the need for greater compatibility with the AAR equipment and procedures of other NATO air forces.

TANKER/TRANSPORT AIRCRAFT IN CURRENT SERVICE

KC-10 Extender

The beginning of the tanker/transport era effectively dates from March 1981 when the USAF took delivery of its first KC-10 Extender.

By any yardstick, the KC-10 is a most impressive aircraft, with a range of capabilities which no other tanker/transport has yet been able to emulate, much less surpass. For example, it is the only in-flight refueller to be equipped with both the boom and hose and drogue systems, reflecting the formal widening of Strategic Air Command's traditional AAR responsibilities to include provision of in-flight refuelling facilities for fighter aircraft operated by Tactical Air Command (TAC), the US Navy and the US Marine Corps. Prior to the advent of the KC-10, these probe-equipped fighters could refuel only from hose-trailing tankers, a number of which were operated by TAC and the US Navy for that reason. However, recognising the importance of exploiting the tanker/transport concept as fully as possible, and thereby optimising its potential for force projection, the Pentagon wisely stipulated that the KC-10 should be configured for both methods of in-flight refuelling.

Accordingly, an independent hose and drogue system is installed at the rear underside of the fuselage, offset to starboard from the boom which as usual is situated in the centre of the rear fuselage (see Plate 3.7). As a result of this dual capability, which enables the aircraft to transfer fuel by either method on a single mission, the KC-10 can refuel:

- All USAF, US Navy and US Marine Corps combat aircraft.
- Virtually all other NATO combat aircraft.
- All US and NATO airlifters equipped for AAR.

Such versatility, added to its considerable airlift capabilities, means that the KC-10 Extender offers substantial flexibility and interoperability, not only in national terms but also in the multinational context.

PLATE 3.6. KC-10 Extender.

PLATE 3.7.   KC-10 with hose and drogue partially deployed.

Mission

As indicated above, the KC-10 force of 60 aircraft was specifically procured to enhance the USA's capacity for rapid worldwide power projection by increasing the mobility and reach of all US forces, particularly those earmarked for contingency operations overseas. The KC-10 contributes to that broad strategic goal by its ability to undertake three important missions, listed here in order of priority:

- Provision of an AAR service during the long-range deployment and redeployment of combat aircraft while simultaneously carrying their support personnel and equipment. This allows air power to be speedily brought to bear wherever it is needed.
- Provision of an AAR service to strategic airlifters such as C-5s, C-141s (and other KC-10s) in order to increase their productivity and reduce their dependence on staging airfields and overflying rights.
- Augmentation of the dedicated strategic airlift force by providing cargo and troop-carrying capacity when required and available.

General Description

A wide-bodied aircraft powered by three General Electric CF6-50C2 turbofan engines, the KC-10 Extender is a military derivative of the McDonnell Douglas DC-10

PLATE 3.8. US Navy F-14 preparing to engage the drogue of a USAF KC-10.

airliner, which is currently in service with some 50 airlines around the world. The USAF's selection of the DC-10 for development into a tanker/transport was based on a careful analysis of the aircraft's performance, conversion potential and price, including both programme and life cycle costs. In order to adapt the basic DC-10 airframe to its new and more demanding military role, a number of major modifications were required, including the installation of additional fuel tanks in the lower cargo compartment, twin AAR systems (flying boom and hose and drogue), boom operator's station and refuelling receptacle. This last was necessary in order to comply with the important principle that tanker/transports should be capable of receiving as well as dispensing fuel.

Importance of AAR Receptacle

The KC-10's ability to refuel from other boom-equipped tankers could be crucial to maintaining the momentum of an airlift or AAR operation, especially when either the Extender itself or the aircraft it is supporting are required to cover exceptionally long sectors with little or no staging facilities *en route*. This capability could also be invaluable in overcoming restrictions on maximum take-off weight at the departure airfield, arising from performance considerations such as the 'WAT limit'[2] and necessitating a reduction in payload in order to carry the required sector fuel. The ability of a strategic airlifter—whether it be the KC-10, C-5 or any other—to trade fuel for payload at take-off and subsequently top up its tanks *en route*, provides air commanders and planners with a most valuable additional option.

PLATE 3.9.   KC-10 preparing to receive fuel from another KC-10 using the flying boom system.

KC-10 Flying Boom System

The KC-10 employs a flying boom of advanced design, specially developed by McDonnell Douglas. It is longer than the KC-135's boom and hence permits greater separation during refuelling. Its fly-by-wire aerofoil system, manipulated by side-sticks at the boom operator's station, provides the high level of controllability which is essential for fast and accurate hook-ups with receiver aircraft. The control system continuously drives the telescopic boom up and down to alleviate any excessive loading induced by changes in the relative positions of tanker and receiver, thus allowing a larger envelope of manoeuvre while reducing the risk of nozzle breakage. Another improvement is the provision of a well-designed station with excellent visibility at the rear of the aircraft which allows the boom operator to sit at a control console, whereas his counterpart in the KC-135 is obliged to lie in an uncomfortable prone position while carrying out his duties. Finally, the KC-10's boom system has been designed to permit an exceptionally high flow-rate, up to a maximum of 1,500 US gallons/min.

KC-10 Performance

AAR Role. The graph in Figure 3.2 highlights the KC-10's impressive performance in the AAR role and gives some indication of its clear superiority over the KC-135.

For example, whereas the KC-135 can transfer only some 110,000 lb of fuel (49,895 kg) at a radius of 1,000 nm, the KC-10 can transfer 200,000 lb (90,719 kg) at a range of 2,000 nm and still return to base. A fuel off-load capability of this order opens up an entirely new range of deployment options—in terms of reach, speed and independence of staging/overflight rights—for combat and airlift aircraft alike.

Airlift Role. The graph in Figure 3.3 not only illustrates the KC-10's impressive

PLATE 3.10. KC-10 boom in stowed position.

capability as a strategic airlifter in its own right but also contrasts its performance with that of the C-5A.

Although the maximum payload of the KC-10, at 170,000 lb (77,112 kg) is less than the 190,000 lb (86,184 kg) that can be carried by the C-5A, the gap in lift capacity steadily reduces at ranges in excess of 4,000 nm until, for sectors between 5,000 and 6,000 nm, there is virtually no difference at all. Comparisons of their respective

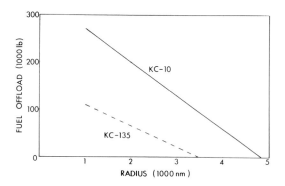

FIG 3.2.   KC-10 performance graph–AAR role.

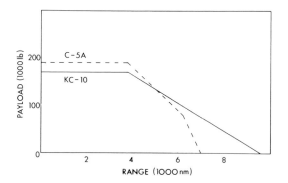

FIG 3.3.   KC-10 performance graph–airlift role.

performance beyond 6,000 nm are largely academic since airlifters tasked to operate over such large distances would normally employ AAR to extend their range with maximum payload. That said, it is sometimes necessary to deploy an airlifter with minimal or even zero payload over exceptionally long sectors. The KC-10's remarkable range in such circumstances was demonstrated in February 1985 when an Extender completed a non-stop, unrefuelled flight of 7,800 nm from Riyadh in Saudi Arabia to March air force base in California in 17.8 hours.

With the whole of the main deck available for the airlift role, another valuable feature of the KC-10 is the range of passenger/cargo permutations that can be accommodated. Options include a maximum of 27 standard '463L' pallets in the all-freight configuration, 25 pallets and 20 personnel, or 17 pallets and 75 personnel. This degree of versatility greatly assists the planning and execution of an airlift operation.

Combined AAR/Airlift Role. By far the most impressive and operationally important aspect of the KC-10's performance rests in its ability to combine both AAR and airlift tasks in a single mission. One of many recent deployments involving the KC-

Fig 3.4. Cutaway diagram of KC-10.

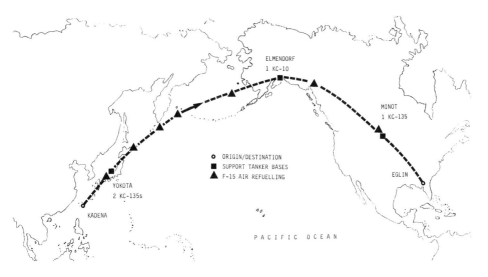

Fig 3.5. Exercise CORONET WEST 1982.

10, illustrated in Figure 3.5, offers a typical example of the aircraft's effectiveness as a tanker/transport.

In this exercise, a single KC-10 was used to support six F-15s on a non-stop deployment from Kadena on Okinawa, Japan to Eglin in Florida, a distance of 7,046 nm. In addition to refuelling the F-15s throughout this long flight, the KC-10 carried 72,000 lb of support equipment and personnel. To accomplish this formidable task, the KC-10 was itself refuelled *en route* by two KC-135s based at Yokota, another KC-10 from Elmendorf in Alaska, and a third KC-135 based at Minot. Even so, the net saving in aircraft was equivalent to eight KC-135s and two C-141s. Not only did two KC-10s thus replace a mix of 10 other tankers and transports, but the mission was accomplished independently of staging airfields in only 14.7 hours.

Summary of Leading Particulars

Crew: Two pilots, one flight engineer, one boom operator, one airloadmaster.
Dimensions: Length   182.0 ft (55.40 m)
                 Wingspan   165.3 ft (50.42 m)
                 Height   58.1 ft (17.70 m).
Engines: Three General Electric CF6-50C2 high bypass ratio turbofans, each producing 52,500 lb of static thrust.
Maximum fuel load: 356,065 lb (161,517 kg).
Maximum payload: 170,000 lb (77,112 kg).
(cargo/passenger)
Maximum take-off weight: 590,000 lb (267,635 kg).
Range with maximum payload: 3,790 nm.
Example of AAR capability: Can transfer 200,000 lb fuel at a radius of 2,000 nm from base.
Cruise speed: 0.82 Mach.
Service ceiling: 42,000 ft.

Tristar

Although its total tanker assets are appreciably less than those of its powerful ally, the UK currently ranks second only to the USA in terms of overall capability. Moreover, the RAF is the only air force apart from the USAF to operate a wide-bodied tanker/transport—in the shape of the Tristar. This is the heaviest aircraft, and is equipped with the most powerful engines, that the RAF has operated to date.

Unlike the KC-10 Extender, the military Tristar is not a derivative but a direct

PLATE 3.11.   RAF Tristar.

conversion of a civilian airliner, in this case the Lockheed L1011-500. The RAF acquired six of these aircraft from British Airways in 1982 and a further three from Pan American Airways in 1984—a total of nine in all. Their conversion from commercial to tanker/transport configuration is being undertaken by Marshall of Cambridge Limited in a major project extending until the early 1990s.

Main Features of Conversion

Conversion of these former airliners to their new role involves a number of important and expensive modifications, including the installation of additional fuel tanks in the two main underfloor baggage holds, and twin hose-drum units (to afford system redundancy) at the aft end of the rear compartment behind the extra tanks. At a later stage of the conversion programme, the RAF plans to fit underwing pods to enable the Tristar to operate as a three-point tanker; accordingly, the new fuel system has been designed to provide a fully integrated tankage so that either of the hose-drum units or underwing pods can be supplied from either the original wing tanks or the new underfloor tanks. Equally, any tank can be filled if and when the aircraft receives fuel through its probe. The probe, which has been provided—in line with basic tanker/transport principles—to confer additional flexibility, protrudes forward over the co-pilot's windscreen. The probe is angled about 7° down from the aircraft datum to compensate for the typical nose-up attitude during refuelling, although this varies with true airpseed. The probe can be removed when not required and replaced by flush-fitting blanking plates, thus affording a small reduction in drag.

Other changes include the strengthening of the fuselage structure underneath the new tanks, because the increased weight exceeds that for which the cargo holds were originally designed; indeed, the weight of the extra fuel in the underfloor tanks alone exceeds the maximum payload of the commercial Tristar. The avionics suite has also been partly modified to bring it more into line with RAF requirements; for example, the global navigation aid OMEGA[3] and IFF equipment have been installed. Apart from these modifications, the major change on the flight deck has been the provision of additional controls and instruments at the flight engineer's station for operation of the AAR systems. External changes include the addition of a pod—containing a closed-circuit television camera—under the rear fuselage and, nearby, twin fairings over the hose and drogue exit tubes.

Tristar Variants

Tristar K Mk 1. This designation is used only for two of the ex-British Airways aircraft intended for use in the tanker/passenger role. To enable the aircraft to undertake its airlift function, the forward section of the main cabin has been modified to incorporate a baggage handling system designed to facilitate the on- and off-load of containerised luggage and small items of freight. The rear two-thirds of the cabin is equipped with seating for 205 passengers.

Tristar K Mk 2. While very similar in appearance and performance to the ex-British Airways aircraft, the RAF's three ex-Pan American Tristars differ in certain important respects which make it impracticable for them to be modified to exactly the same standard as the K Mk 1. For example, differences in the design of the lower rear fuselage are likely to reduce the fuel capacity of the ex-Pan American aircraft by some 10,000 lb (4,540 kg) compared with the K Mk 1. To distinguish them from the K Mk 1, both for operational and maintenance purposes, these three aircraft are designated K Mk 2. Like the K Mk 1, however, the K Mk 2 will operate only in the tanker/passenger role.

Tristar KC Mk 1. The remaining four ex-British Airways aircraft are designated KC Mk 1 and modified for use in the tanker/cargo/passenger role. The KC Mk 1 has the same AAR systems as the K Mk 1 but, to allow the airlift of large and heavy items of freight, is equipped with a cargo door, strengthened floor and integral pallet handling system. In terms of volumetric capacity, the aircraft can carry a maximum of 20 standard pallets (108 × 88 in or 2.74 × 2.23 m). Four winches are installed— two at each end of the cabin—to help with loading and unloading. A fifth winch mounting is provided opposite the cargo door; any of the four winches can be temporarily fitted in this location to pull a load into the aircraft from the external loading platform. In terms of weight, the KC Mk 1 can carry up to 100,000 lb of cargo or fuel (see Summary of Leading Particulars). In the passenger role, up to 195 passengers can be carried on palletised seats, with their baggage stowed on other pallets. To permit operations in support of combat aircraft deployments—the classic tanker/transport mission—the KC Mk 1 can be configured to carry various combinations of passengers, freight and fuel.

Mission

In its new military role, the Tristar's mission can be classified under three headings:

Tanker Operations. Having been acquired to expand the RAF's existing tanker force, the Tristar is primarily a tanker, its main task—especially in war—being the provision of AAR support for fighters and AEW aircraft operating in the UK Air Defence Region.

Tanker/Transport Operations. In line with the RAF's ongoing need to deploy combat aircraft over long distances (often outside the NATO area) for operations, exercises and training, the Tristar is also required to provide the simultaneous AAR and airlift support needed for the expeditious execution of such deployments.

Transport Operations. The Tristar's combination of range, payload and speed make it a most efficient and cost-effective strategic airlifter of both cargo and personnel. Whilst it is unlikely that the aircraft could be spared for this role in war, when it would be needed for the tasks outlined above, it is extensively employed on peacetime transport operations, notably in the South Atlantic in support of the Falklands garrison.

Performance

For normal operations, the RAF's Tristar has a maximum all-up weight of 525,000 lb compared with the 504,000 lb of the airline version. This increase, needed to allow the aircraft to uplift its maximum fuel load, is achieved by reducing the in-flight 'g' limit from 2.5 to 2.0. Since there has been no change in the original RB211 engines, take-off performance at this higher gross weight is somewhat reduced, thereby increasing field length requirements to more than 10,000 ft (at sea level in standard atmospheric conditions). This means that, quite apart from any load-bearing considerations, there are relatively few RAF airfields from which a fully laden Tristar can operate within normal safety criteria. Hence, in the absence of a suitable runway, or where the WAT limit imposes other constraints on weight, a Tristar requiring maximum fuel for a particular mission will need to top up its tanks after take-off or whilst *en route*.

In the purely transport role, with the underfloor tanks empty, the Tristar will probably not need to operate at weights above 510,000 lb; with its maximum payload of 100,000 lb and using wing fuel only, the aircraft has a range of 3,000 nm. In the tanker role, it can transfer 124,000 lb of fuel at a radius of 2,300 nm from base. In the tanker/transport role, the Tristar KC Mk 1 can provide AAR support for four combat aircraft and simultaneously carry 20,000 lb of payload over a sector of 3,000 nm.

It is anticipated that the Tristar will eventually be cleared to operate as a tanker in the speed range 180–320 kt IAS, thereby enabling it to refuel any probe-equipped fixed-wing aircraft. The 'toboggan' technique may be necessary in some cases, especially when the C-130 Hercules is the receiver. This procedure is used not because these aircraft cannot formate at the same slow speed but because downwash from the Tristar tends to push the Hercules two or three degrees below the horizontal, forcing the latter's pilot to compensate with additional power—equivalent to a rate of climb of 800–1,000 ft/min—just to maintain straight and level flight. Using the 'toboggan' technique, tanker and receiver make initial contact at about 25,000 ft and

PLATE 3.12. Tristar refuelling an RAF Phantom.

then gradually descend to about 8,000 ft during the transfer of fuel. This allows the receiver to accept a steadily increasing all-up weight without having to apply excessive climb power throughout the AAR process.

Advanced Features

Although in service since 1979 as the L-1011-500 airliner before its acquisition and conversion by the RAF for the tanker/transport role, the Tristar is still one of the most advanced large aircraft in military use. The more important of its special features are outlined below:

Flying Tail. Instead of the conventional horizontal tail system with stabiliser and elevator, the Tristar has a 'flying' tail which confers a wider margin of longitudinal control, especially at critical speeds. For example, the flying tail allows full control to be maintained regardless of trim position, thereby eliminating the dangers involved in mis-trimming, stalled trim or runaway trim.

Direct Lift Control. Conventional lift control acts indirectly; a change in pitch altitude is required to change the lift. The Tristar, however, has a direct lift control system operated through wing-mounted and fast-acting spoilers which alter the aerofoil shape—and hence the lift which it is generating—at a constant pitch altitude. By affording better control of vertical speed during approach and landing, this system allows the pilot to fly down the glideslope with great precision, to flare the aircraft at a lower height and to achieve a more accurate touchdown. It also allows him to take corrective action more rapidly and effectively if wind shear is encountered.

Performance Management System. A problem common to all modern jet transports is the difficulty of maintaining constant airspeed to achieve the optimum cruise regime, a task which can require virtually continuous adjustment of the throttles. The Tristar's performance management system was designed to overcome this problem by effecting minor changes in speed through automatic variations in pitch, reverting to throttle adjustment only if a ± 50 ft altitude tolerance is exceeded.

Quadruple Hydraulics. The Tristar has four hydraulic systems, each pressurised by an engine-driven pump and each provided with a secondary and independent source of pressure. All primary and secondary flight controls have multiple sources of hydraulic power, so arranged that no failure will cause asymmetry. Indeed, the degree of redundancy is such that the aircraft can still be flown safely if three of the four hydraulic systems fail.

Propulsion. The Tristar is powered by three Rolls-Royce RB211-524B engines incorporating advanced high bypass ratio technology. These engines feature a three-shaft design (unique to Rolls-Royce) which separates the fan and its turbine from the gas generator core; this improves the internal aerodynamics, resulting in greater compressor stall margins that permit rapid thrust changes without surging. The three-shaft design also allows a reduction in the number of stages, producing a shorter, more rigid engine which affords better control of blade tip clearances. Another important feature of the RB211 engine is its modular construction which greatly facilitates maintenance and rectification.

Reverse Thrust System. The RB211's thrust reverser is attached to the basic engine, thereby producing a direct load vector which avoids the problem of relative movement that sometimes occurs with airframe-mounted reversers. Moreover, the relatively low thrust line of the centre engine—compared with some other tri-jets—prevents nosewheel lift-off when reverse thrust is applied.

Active Aileron Controls. Aircraft designers have long known that increased wingspan, with its higher aspect ratio, reduces drag and hence improves fuel consumption. However, the benefits of lengthening the wings of existing aircraft have hitherto been far outweighed by the cost involved in redesigning and strengthening the basic wing to compensate for the increase in structural loading, a process also entailing a punitive addition to the basic weight. In developing the L-1011-500 Tristar from the earlier family of L-1011 airliners, which had a wing span of only 155 ft 4 in, Lockheed designers—seeking improved fuel economy—were able to increase the wing span by 9 ft to 164 ft 4 in without incurring such penalties. They achieved this by designing an 'active aileron control' system which automatically applies load-relieving inputs to the outboard ailerons provided on the wing-tip extensions. As illustrated in Figure 3.6, when vertical accelerometers sense an increase in wing loading due, for example, to manoeuvres or wind gusts, both ailerons are deflected upwards, thereby reducing lift over the outer wings. This redistributes total lift inboard, effectively reducing the flexing moments to the level of the original wing structure.

Summary of Leading Particulars (KC Mk 1)

Crew: Two pilots, one flight engineer and one airloadmaster (plus cabin crew according to number of passengers carried).

Dimensions   Length   164.2 ft (50.0 m)
                 Wingspan   164.3 ft (50.1 m)
                 Height   55.3 ft (16.9 m).

Engines: Three Rolls-Royce RB211-524B turbofans each producing 50,000 lb of static thrust.

Maximum take-off weight: 525,000 lb (238,135 kg).

Maximum fuel capacity: 300,000 lb (136,077 kg).
(all transferable)

Maximum no of passengers: 195.

ACTIVE AILERON CONTROLS

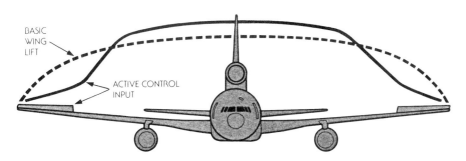

FIG 3.6.   Tristar active aileron control system.

Maximum payload:* 100,000 lb (45,359 kg).
Maximum range with maximum payload: 3,000 nm.
Example of AAR capability: Can transfer 124,000 lb (56,245 kg) of fuel at a radius of 2,300 nm from base.
Cruise speed: 0.83 Mach.
Service ceiling: 43,000 ft

*The maximum load that can be carried in the fuselage is 100,000 lb, which *includes* any fuel carried in the lower hold tanks, which themselves hold 100,000 lb of fuel. Thus any fuel carried in the fuselage underfloor tanks reduces the weight of payload by the same amount.

TANKER/TRANSPORT CAPABILITIES OF OTHER AIR FORCES

As far as can be ascertained from open sources, neither the USSR nor any of its Warsaw Pact allies currently operates a tanker/transport aircraft as defined earlier in this chapter. Indeed, Western observers believe that the wider concept of AAR is still at a relatively early stage of development in the USSR compared with the USA, UK, France and various other countries. Whilst a number of Soviet Bison and Badger aircraft appear to have been converted into tankers, it is thought that they are intended to provide AAR support for long-range bomber and maritime patrol aircraft; there is no evidence to suggest that Soviet fighters, much less strategic transports, are equipped for in-flight refuelling. In view of the Kremlin's readiness to pour money into military projects, it would seem that the tanker/transport concept does not yet feature in Soviet strategy. Presumably this is because, to date, the USSR has seldom needed to deploy combat aircraft as rapidly as possible over strategic distances beyond its own boundaries, preferring instead to exert pressure by using its considerable naval power and straightforward airlift resources. Whether it can afford to forego the advantages of the tanker/transport in the longer term remains to be seen.

Elsewhere in the world, while the merits of AAR are gaining increasing recognition, few nations have either the need or resources to operate aircraft such as the KC-10 or Tristar. Some air forces have instead opted for a more limited tanker/transport capability by operating multi-purpose aircraft which can be rapidly switched from AAR to transport or other tasks (and vice versa) but which offer at best only a modest ability to perform AAR and airlift roles simultaneously. The leading aircraft in this category is a modified version of the B-707-320 airliner—redesignated the KC-137—which the Boeing Military Airplane Company is now offering as either a two- or three-point tanker while retaining a useful transport capability. The Brazilian Air Force took delivery of the first of four of these aircraft in November 1986, opting for a two-point tanker with hose and drogue pods under each wing. The KC-137 is not only an effective tanker but can also be readily converted to all-passenger, all-cargo or mixed passenger/cargo configurations. Passengers and cargo *can* be carried during AAR missions but the KC-137's capacity for simultaneous tanker/transport operations is very modest compared with that of the KC-10 and Tristar. The chief advantage of the KC-137 lies not in its capacity for simultaneous tasking, but in its ability to undertake a wide range of missions—such as VIP, passenger, freight, EW and airborne command post—when not needed as a tanker. It is the prospect of such cost-effective versatility that has led Spain to order two KC-137s, and Australia to seek tenders for the conversion of its four existing 707 transports to KC-137 standard. Recent reports also suggest that, for similar reasons, the South African Air Force plans to form a squadron of four modified 707s with tanker/transport capability. Multi-purpose aircraft such as the KC-137 certainly offer worthwhile flexibility and economy; but they cannot provide the strategic potential of the fully fledged tanker/transport.

Questions

1. (a) What is the essential characteristic of a tanker/transport aircraft as defined in this Chapter?

    (b) Describe the classic tanker/transport mission.

2. Explain briefly how AAR can be used as an 'extender' and 'multiplier' of air transport forces.

3. (a) Name the two basic AAR systems.

    (b) Compare the respective advantages and disadvantages of each of these systems.

4. Describe the 'toboggan' technique sometimes employed in AAR.

5. Why is it important that a tanker/transport aircraft should be able to receive as well as dispense fuel in flight?

6. Describe three advanced features of the KC-10's fuel-dispensing system which contribute to that aircraft's superiority over the KC-135.

7. In what important respects does the RAF Tristar KCMk1 differ from the KMk1 and KMk2?

8. The RAF Tristar is equipped with 'active aileron controls'. What is their primary function?

4

Tactical Operations

Just as strategic airlift operations play a crucial role in the rapid projection of power over intercontinental distances, so tactical airlift operations make a vital contribution to the application of military force within a theatre or defined geographical area. As discussed in Chapter 2, it is sometimes difficult to draw a clear distinction between the two classifications. When, for example, a long-range flight terminates in a low-level phase and para-drop, this can either be represented as a strategic mission with a tactical dimension or vice versa. The question is further blurred because, while some strategic airlifters have a tactical capability, there are also some tactical transports such as the C-130 which have certain inter-theatre qualities and applications. That said, most of the current generation of tactical airlifters tend not to have the same range, speed and cargo capacity as the larger strategic aircraft. Nevertheless, the ability of a tactical airlifter (whether fixed or rotary wing) to enhance a state's overall military potential and posture is widely recognised. Hence, whilst only those states with intercontinental responsibilities or aspirations—and a budget to match—need or can afford strategic airlift capability, almost every state that has an air force can boast at least some tactical transport capability.

As indicated in the outline of roles in Chapter 1, tactical transport aircraft are

PLATE 4.1. Aeritalia G-222 of 46 Air Brigade, Italian Air Force.

employed on a wide variety of missions covering the whole spectrum of deployment, resupply, aeromedical evacuation and redeployment. Exploiting the classical characteristics of air power, tactical aircraft can operate with speed and flexibility in support of combat forces, virtually irrespective of terrain and if necessary using only semi-prepared airstrips. Naturally, not all tactical airlifters can undertake the whole range of tasks although there are those—notably the C-130—whose versatility allows them to perform virtually every tactical transport mission from full-scale parachute assault to routine logistic or troop-carrying flights. Similarly, some types of support helicopters—covered separately in Chapter 7—are more capable than others. Nevertheless, there are common qualities which all tactical airlifters should possess if they are to perform well in their exacting role. The more important of these are outlined below.

AIRCRAFT REQUIREMENTS

Aircraft designed for tactical transport operations should satisfy the following basic requirements:

- Rugged construction to withstand the general wear and tear of the tactical airlift role.
- Short Take Off and Landing (STOL) capability at both paved and unpaved airfields.
- Good payload, with cargo compartment capable of carrying the largest items of airportable equipment used in the theatre of operations.
- Rear cargo doors and integral ramps to facilitate rapid loading and unloading.
- Rapid reconfiguration from one role to another (eg, from troop-carrying to airlift of palletised supplies).
- Air-drop capability.

Ideally, such aircraft should also have:

- An all-weather capability.
- Good slow-flying characteristics.
- Radius of action of about 1,000 nm with full payload, and ferry range of about 3,000 nm.
- The capability to receive fuel in flight.
- Reliable systems designed for minimum off-base maintenance.

In design terms, many of these requirements are conflicting; for instance, heavy payloads are difficult to reconcile with STOL operations. Hence, in order to strike an optimum balance between the various criteria outlined above, the aircraft designer is usually forced to compromise. In practice, this leaves him with little room for manoeuvre and he generally opts for a design which incorporates a high wing with a low-slung fuselage, a feature which not only helps to protect the engines and flaps from Foreign Object Damage (FOD) during operations from unpaved surfaces, but also maximises the cross-section and volumetric capacity of the cargo compartment. Almost always, this basic configuration is complemented by an upswept tail which permits the provision of aft cargo doors and a rear ramp, the ramp being necessary for the air-dropping of equipment as well as for easy cargo handling on the ground. Having provided the necessary space and access/egress for payload, the designer must also equip the aircraft with high-lift and slow-flying devices which, combined with a robust and high-flotation landing gear in side fairings, are essential if the aircraft is to be used for STOL or quasi-STOL operations from unprepared surfaces. Finally, the designer must provide engines of suitable power and reliable, easy-to-maintain systems which will allow sustained operations from austere airfields with minimal servicing and rectification facilities. The net result is a generation of tactical airlifters that tend to be variations on a basic theme. This will be clear from the following section which examines the more important types currently in service; for example, of the six leading tactical transports surveyed, only the F-27 Friendship (see Plate 4.14) does not have rear cargo doors. Although the advent of new technology and materials offers scope for significant improvement in performance and capability, most tactical transports in operational service today were designed and built before these were available. However, as with the new generation of strategic airlifters, designers of future tactical airlifters should be able to exploit new technology to telling effect.

TACTICAL TRANSPORT AIRCRAFT IN CURRENT SERVICE

There are many different types of tactical transports currently in operational service throughout the world; this section is restricted to an examination of a representative cross-section of large, medium and small airlifters, beginning with the C-130—the aircraft which more than any other has dominated this aspect of air power during the past three decades.

C-130 Hercules

Few would dispute the proposition that, like the remarkable Douglas Dakota DC-3, Lockheed's C-130 Hercules is one of the most effective and successful airlifters ever built. Many adjectives and superlatives have been used to describe this truly ubiquitous aircraft, but nothing testifies more eloquently to its special qualities than the duration and scale of its production run and the extent of its worldwide sales. Since entering operational service with the USAF in 1957, over 1,850 military C-130s and commercial derivatives have been produced for some 57 air forces and 31 civilian freight carriers. With Lockheed's production line at Marietta, Georgia, still turning out three aircraft per month, and with further orders expected for several years to come, there is every chance that the eventual total will reach 2,000 aircraft. Likewise, with an average life expectancy of about 35 years, the C-130 is certain to remain in front-line service well into the next century.

Naturally, the C-130 rolling off today's production line bears only a superficial resemblance to the early models of 30 years ago. Throughout this period, the aircraft has been under continuous development, partly in response to emerging requirements by users and partly at the manufacturer's initiative in order to take advantage of new technology and components. Improvements over the years have included more powerful and fuel-efficient engines, a better auxiliary power unit and more efficient

PLATE 4.2. A C-130 of the USAF delivering an armoured vehicle by the Low Altitude Parachute Extraction System (LAPES). This technique is described in Chapter 6.

brakes. In addition, virtually every system has been modernised; the fuselage skin, wing panels, wing structure, landing gear, propellers, hydraulic systems, fuel system, avionics and electrical systems have all been systematically reviewed and updated.

With these changes came new marques, the C-130A giving way to the C-130B in 1959, in turn replaced in 1962 by the C-130E which was itself superseded by the C-130H model in 1964. The most radical development was the introduction of a stretched version (the C-130H-30, illustrated in Plate 4.3) in 1980, available either as a new production aircraft or as a modification to C-130H aircraft already in service. The main advantage of this new variant—examined with the other models in more detail below—was to increase cargo capacity by some 30% without diminishing the aircraft's performance or tactical versatility.

As a result of this continuous evolutionary process, the current production model has more power, increased fuel capacity and greater range than earlier versions, whilst offering greater reliability and needing fewer maintenance man-hours per flying hour. Above all, stretched and unstretched variants alike can carry more payload, thereby reinforcing the C-130's reputation for setting the standards by which all other tactical airlifters are currently judged.

Multi-role Capability

The primary mission of the C-130 is to transport personnel or cargo either by air-landing or air-dropping, operating if required from short, unprepared airstrips. Nevertheless, it is important to note that not all C-130s are employed in the airlift role. On the contrary, one of the main reasons for the aircraft's international

PLATE 4.3.   A stretched C-130 of the RAF (designated Hercules C Mk 3). All stretched C-130s in service with the RAF are equipped for in-flight refuelling.

popularity and its exceptionally long production run has been its remarkable adaptability to a wide range of other missions. To cite just a sample of its many applications, variants of the C-130 have variously been used as tankers, gunships, minelayers and airborne command posts as well as for search and rescue, weather reconnaissance and aerial photography. This versatility has made the Hercules one of the most impressive multi-role aircraft in aviation history, especially as many of these special derivatives are readily convertible from and to aircraft that can otherwise be employed primarily as airlifters. Such flexibility offers both operational and financial advantages: senior air commanders can readily switch the aircraft from one type of mission to another as the situation demands, while reaping savings in procurement and operating costs by minimising the number of aircraft types in the air force inventory. With defence budgets everywhere under increasing pressure, the trend towards multi-mission capability—already well established as an essential principle of modern air power—is likely to become even more pronounced in the future. By virtue of their basic configuration, offering a combination of large payload capacity and endurance, airlifters such as the C-130 are particularly suited for adaptation to a variety of non-transport missions while retaining their fundamental airlift capabilities. The relevance of this concept of wider employability to future developments is examined in more detail in Chapter 8.

General Description of C-130

There are four basic versions of the military Hercules: the C-130A, C-130B, C-130E and C-130H. Before comparing their individual characteristics, it is worth noting those features which are common to all of these models. All versions have a spacious, fully pressurised cargo compartment with rapid loading and unloading via full-width rear doors and integral ramp, the ramp being fully adjustable between ground level and truck-bed height. The floor of the cargo compartment is strengthened to accept heavy loads, with D-ring tiedown points being provided on the walls as well as on the floor and ramp. Special fittings are also provided to permit the rapid installation of stretchers and troop seats. Side-mounted seats, stowed in the folded position when not required, are normally carried on all flights. A removable rail-roller conveyor system, extending the full length of the cargo floor, can be fitted to allow containers or pallets (up to 9 ft 10 in wide) to slide directly from aircraft hold (10 ft 2 in wide) to transporter or vice versa. (See Plates 4.4 and 4.5.)

All four basic models have the same external and internal dimensions (see Table 4.1) but differ in other important respects (see Table 4.2).

C-130A. Although it is over 30 years since the C-130A first entered operational service, over 100 are still flying. All have been modified to incorporate four-bladed propellers in place of the original three-bladed versions and most have been fitted with a 450-gallon (US) pylon-mounted fuel tank under each wing to increase range or endurance.

C-130B. With a strengthened airframe, increased fuel capacity and more powerful engines, the C-130B has a better range and performance than the C-130A. (See Table 4.2.) The original Allison T56-A-7 engines were later replaced on all US Navy and Marine Corps aircraft by the more powerful T56-A-15, the standard power unit on the C-130H. Over 120 C-130Bs are still in service with the US Armed Forces and other air forces around the world.

C-130E. By further strengthening the basic construction and adding a 1,360-gallon (US) pylon-mounted fuel tank under each wing, Lockheed not only increased the

PLATE 4.4.   C-130 roller conveyor system.

PLATE 4.5.   Large container ready for direct loading from truck to C-130.

TABLE 4.1   *External/Internal Dimensions of*
C-130A/B/E/H

External	
Wing span	132.6 ft
Length	97.8 ft
Height	38.1 ft
Cargo hold	
Length	40.1 ft
Width	10.2 ft
Height	9.1 ft
Usable volume	4,500 cu ft

TABLE 4.2   *Comparison of C-130A/B/E/H Performance Data*

	Unit	C-130A	C-130B	C-130E	C-130H
Engines					
Type	–	T56-A-9	T56-A-7	T56-A-7	T56-A-15
Take-off power	eshp	3,750	4,050	4,050	4,508
No of propeller blades	–	3/4	4	4	4
Propeller diameter	ft	15/13.5	13.5	13.5	13.5
Operating Weights					
Maximum gross weight (2.5 g)	lb	124,200	135,000	155,000	155,000
Maximum take-off weight (2.5 g)	lb	124,200	135,000	155,000	155,000
Maximum landing weight (5 fps)	lb	124,200	135,000	155,000	155,000
Maximum landing weight (9 fps)	lb	96,000	118,000	130,000	130,000
Maximum payload (2.5 g)	lb	35,000	35,000	45,000	45,000
Fuel capacity	lb	39,975	45,240	62,920	62,920
Maximum range with maximum payload	nm	1,900	2,100	2,100	2,100
Maximum ferry range	nm	2,900	3,900	4,500	4,500

C-130E's maximum gross weight and payload but also conferred on this model the range needed for strategic missions. Whereas the C-130B can carry a payload of 30,000 lb over a distance of only 2,400 nm, the C-130E can airlift the same payload over a sector of 3,100 nm. Although intended primarily for longer-range missions, the C-130E—over 400 of which are still in operational service around the world— retains all of the tactical attributes of the earlier model.

C-130H. The C-130H, Lockheed's current production model, features improved systems, state-of-the-art avionics and more powerful engines, described in more detail below. Fire suppression foam built into the fuel tanks of USAF C-130H aircraft provides additional fire protection but reduces total fuel capacity by about 4.5%; other H models are similar to the C-130E in terms of payload and range. Well over 400 versions of the C-130H are currently in military service worldwide, including the USAF's MC-130H used for operations by special forces.

- Aircrew:
  The typical crew comprises two pilots, a navigator, a flight engineer and a loadmaster. The flight deck is spacious (see Plate 4.6) with nearly 40 sq ft of window providing excellent all-round visibility, a particularly important feature in tactical operations.

- Avionics:
  The various avionic sub-systems have been continuously updated over the years to take advantage of developing technology. Although the combination of equipment varies widely according to the specifications and requirements of individual operators, the typical suite (illustrated in Table 4.3) is extremely comprehensive. In addition to using conventional navigation aids such as VOR and TACAN[1] (when within range of the relevant ground-based installations)

PLATE 4.6.   C-130 flight deck.

PLATE 4.7.   Pilots' instrument panels on C-130.

TABLE 4.3   *A typical Hercules Avionics Suite*

Hercules Avionics

Communication
>   Intercommunication System (AN/AIC-18)
>   Public Address System (AN/AIC-13)
>   HF Communication Radio (Dual) (628T-1)
>   VHF Communication Radio (Dual) 618M-3A)
>   UHF Communication Radio (AN/ARC-164)
>   ATC Transponder (Dual) (621A-6A)

Navigation
>   UHF Direction Finder (DF-301E)
>   Automatic Direction Finder (Dual) (DF-206)
>   Marker Beacon Receiver (51Z-4)
>   VHF Navigation System (Dual) (51RV-4B)
>   DME (Dual) (860E-5)
>   Radio Altimeter (AL-101)
>   Weather Radar (RDR-1F)
>   OMEGA Navigation System (CMA 771)
>   Inertial Navigation System (LTN-72)
>   Compass System (Dual) (C-12)
>   Ground Proximity Warning System (MK II)

Flight Controls
>   Autopilot (AP-105V)
>   Flight Director System (Dual) (FD 109)

many C-130s are equipped with OMEGA for transoceanic or 'off-airways' flights. OMEGA is a very good aid which allows rapid and accurate determination of the aircraft's position in most areas of the world, but it depends on signals transmitted by a number of ground stations[2]. The stations can become unserviceable in peacetime and some may cease to function altogether in war. Whilst it too can suffer unserviceability, an INS has the advantage of being a self-contained piece of equipment which can provide a continuous and highly accurate supply of navigational data independently of any external references. Moreover, being a passive system which does not involve the transmission or reception of radio or radar signals, INS is extremely valuable in tactical transport operations where the use of other aids might increase the risk of electronic detection. With safety of paramount importance in peacetime transport operations, Lockheed's latest avionics package includes an improved colour-scope weather-avoidance radar (with ground mapping mode); a radio altimeter to indicate the aircraft's absolute altitude (especially useful for low-level operations); and a ground proximity warning system to alert the crew when the aircraft is too low for the prevailing configuration and flight regime.

- Engines:
  The C-130H is powered by four Allison T56-A-15 propjets. They are capable of generating 4,910 equivalent shaft horse power (eshp) at sea-level in standard atmospheric conditions but in practice are restricted to 4,508 eshp in order not to exceed nacelle/wing structural limitations. The T56-A-15 has a lower specific fuel consumption yet produces some 11% more power than the 'Dash 7' engine used on earlier models, permitting higher cruise speeds and altitudes as well as shorter field length requirements. The engine has a four-bladed, variable pitch, fully reversible propeller unit. In flight, the propellers operate at constant speed, providing an instantaneous response to control inputs from the flight deck. On the ground, the aircraft's reverse thrust capability not only assists landing performance but also enhances its parking manoeuvrability by allowing it to back up under its own power.

- Landing Gear/Unpaved Surface Operations:
  The C-130's ability to operate from a wide variety of unpaved surfaces (including grass, gravel, dirt, sand and steel-matting) is one of its most important tactical characteristics, greatly increasing the number of airstrips into which troops and equipment can be deployed. The aircraft's capacity to absorb the heavy shocks and stresses imposed by such operations derives from its heavy-duty landing gear, comprising four mainwheels (two in tandem on each side) and twin nosewheels. The main gear is so designed that the large diameter, wide, low-pressure tyres spread the wheel loads over a large contact area, thus affording excellent flotation qualities, with the tandem arrangement allowing the front tyres to make and pack a path for the rear tyres on unpaved surfaces. When combined with reversible engine thrust, the steerable nose gear assists manoeuvrability on the ground, particularly when the wheels become temporarily stuck in mud, soft sand or loose gravel. For extra structural efficiency, the main gear is housed in wells directly below the fuselage.

- Payload:
  The C-130H can carry many permutations of payload as illustrated in Table 4.4.

Stretched Hercules (C-130H-30)

Also still in production, the C-130H-30 is an extended-fuselage version of the basic C-130H, stretched by the insertion of two plugs of 100 in and 80 in forward and aft of the wing respectively. This increases the overall length of the aircraft and hence the cargo compartment by 15 ft, thereby boosting the compartment's cubic capacity by some 30% and permitting the loading of two additional standard pallets. Although the modification adds 3,773 lb to the aircraft's basic weight, thus reducing its maximum payload by the same amount (to 38,900 lb compared with the C-130H's maximum payload of 42,673 lb), there is negligible effect on airfield or cruise performance. The modest weight penalty is more than offset by the C-130H-30's ability to carry bulkier loads than unstretched models, which often run out of volumetric capacity before reaching their maximum payload weight. There are also other advantages. For example, by stretching 30 of their fleet of 62 C-130s, the RAF effectively gained nine additional aircraft in terms of cubic capacity without any need for extra aircrew or maintenance personnel. Moreover, far from impairing the overall tactical capability of their Hercules Force, the RAF's acquisition of the stretched version enhanced its tactical potential by virtue of the larger model's ability to air-drop 92 paratroops *vis-à-vis* the unmodified aircraft's 64. The RAF's unstretched and stretched Hercules aircraft, respectively designated C Mk 1 and C Mk 3, are pictured in Plate 4.8; their vital statistics are compared in Figure 4.1.

TABLE 4.4   *C-130H – Typical Payloads*

Air landing role	Air-drop role
92 troops	64 paratroops
74 stretchers and 2 attendants	16 × 1-ton containers
16 × 1-ton containers	2 × medium-size platforms
3 × Land-Rovers/trailers plus 14 troops	
2 × 4-ton trucks plus 15 troops	
1 × Puma helicopter plus 10 troops	
2 × Complete Rapier missile systems plus 60 missiles	

AN-12 Cub

Often described as the Soviet Hercules, the Antonov AN-12 (code-named Cub by NATO) dates from the same era as its American counterpart, having entered service with the *VTA* in 1959. Like the C-130, the AN-12 was designed for commercial as well as military use; just over 900 aircraft were built in all before production ceased in 1973. By then, the AN-12 (designated AN-12BP in the USSR) had become firmly established as the standard Soviet tactical transport, but the following year saw the beginning of a programme—still under way today—to replace it with the more capable IL-76 (described in Chapter 2). As far as can be ascertained from unclassified sources, some 250 AN-12s currently remain in service with the *VTA*, while another 200 or so continue to fly in Aeroflot livery. The Soviet state airline and air force

PLATE 4.8. RAF Hercules C Mk 1 (unstretched) and C Mk 3 (stretched).

have always been the main operators of the AN-12 although the aircraft has also been successfully operated by the air forces of China,[3] Czechoslovakia, India, Iraq, Poland and Yugoslavia. As AN-12s have been phased out of *VTA* service to be superseded by the IL-76, the Soviet authorities have as usual assigned some of the AN-12s to other tasks rather than retire them abruptly from operational service. Hence a large number of former *VTA* AN-12s are now employed by the Soviet Air

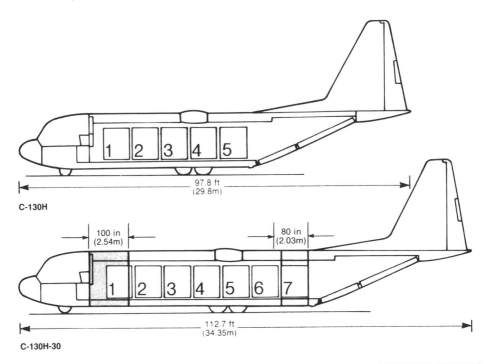

HERCULES MODEL	CARGO LENGTH	CARGO WIDTH	CARGO HEIGHT	RAMP LENGTH	NO. 463L PALLETS	RAMP* PALLETS	LITTERS	COMBAT TROOPS	PARA TROOPS	VOLUME
C-130H	40.1 ft 12.2 m	10.2 ft. 3.1 m	9.1 ft. 2.8 m	10.3 ft 3.1 m	5	1	74	92	64	4,500 cu ft 127.4 cu m
C-130H-30	55.1 ft 16.8 m	10.2 ft. 3.1 m	9.1 ft. 2.8 m	10.3 ft 3.1 m	7	1	97	128	92	5,845 cu ft 165.5 cu m

*Ramp pallet utilizes ramp space

FIG 4.1. Comparison of C-130H and C-130H-30.

Force and Navy in the Electronic Intelligence Gathering (ELINT) and Electronic Warfare (EW) roles. Although also used as a moderately effective bomber by the Indian Air Force, the AN-12 has not proved nearly so adaptable to non-transport missions as the C-130.

While superficially similar in appearance to the Hercules, the AN-12 has a number of distinctive features, as can be seen in Plates 4.9 and 4.10. The most striking differences occur at the aircraft's extremities, the AN-12 having a glazed nose, anhedral (ie, downward-displaced) wing tips—as opposed to the C-130's dihedral wings—and a tail turret containing twin 23-mm cannon. The rationale for the cannon is difficult to understand since the guns are manually laid and have a limited arc of fire; if crewed by a tail gunner during an opposed airborne assault operation, this token armament would be unlikely to pose much of a threat to enemy air or ground forces. Externally, the AN-12 is some 10 ft longer than the unstretched C-130, but its cargo hold is only 4 ft longer than that of the American aircraft. Despite these greater dimensions, the usable volume within the AN-12's hold is only 3,433 cu ft compared with the 4,500 cu ft available in the C-130. Moreover, the AN-12's hold is unpressurised;

PLATE 4.9.   AN-12 Cub of Egyptian Air Force.

this means that, if passengers are carried, the aircraft cannot fly above 10,000 ft, a restriction which significantly increases fuel consumption and hence reduces range. Cargo is loaded through large doors under the upswept rear fuselage which hinge upwards and inwards to allow vehicles to transload freight directly from truck-bed height into the hold. However, the absence of an integral rear ramp necessitates the carriage of portable equipment (or its provision at airfields) if vehicles are to be rapidly on or off-loaded. The AN-12 also compares unfavourably with the C-130 in performance terms. With less powerful engines, its maximum take-off weight is 20,000 lb less than that of the C-130H and, although it can carry a similar payload, its range is restricted by its more limited fuel capacity. The C-130H can in fact carry twice as much fuel as the standard AN-12.

Summary of Leading Particulars

Flight deck crew: Two pilots, navigator, flight engineer and radio operator. (The navigator's station is in the glazed nose beneath the flight deck.)

External dimensions:	Length	108.6 ft (33.10 m)
	Wingspan	124.7 ft (38.00 m)
	Height	34.5 ft (10.53 m).
Cargo hold dimensions:	Length	44.3 ft (13.50 m)
(excluding ramps)	Maximum width	11.5 ft (3.50 m)
	Maximum height	8.5 ft (2.60 m)
	Volume	3,433 cu ft (97.20 m³).

Engines: Four Ivchenko AI-20K turboprops, driving four-bladed reversible-pitch propellers. Each engine delivers 4,000 eshp.

Basic weight: 61,730 lb (28,000 kg).

Maximum payload: 44,090 lb (20,000 kg).

Passenger capacity: Air-land—90 troops

Air-drop—60 paratroops (despatched via rear doors).

Maximum fuel: 30,000 lb (13,600 kg).

PLATE 4.10.　AN-12 of Aeroflot.

Maximum take-off weight: 134,480 lb (61,000 kg).
Range with maximum payload: 1,942 nm.
Ferry range (maximum fuel): 3,075 nm.
Cruise speed: 305 kt.
Service ceiling: 33,500 ft.
Typical take-off distance: 2,460 ft. (750 m).
Typical landing distance: 1,800 ft (550 m).

C-160 Transall

The C-160 Transall derives its name from the Transporter Allianz, a consortium of French and West German aerospace companies formed in 1959 to design, develop and produce a twin-turboprop, general-purpose, tactical airlifter for the air forces of France and the Federal Republic of Germany. Following the maiden flight of the first prototype in 1963, deliveries began in 1967 by which time the air forces of Turkey and South Africa had also placed orders for the C-160. The original production line closed in 1972, but in response to a new order from France, resumed at Toulouse in 1977 with a further 35 aircraft being built before the line finally closed down in 1986.

The French Air Force received 29 of the second series C-160 which, apart from the addition of a flight refuelling probe above the flight deck, is outwardly almost identical to the previous model. However, it has better avionics (see Figure 4.2) and— thanks to the installation of an additional fuel tank in the centre section of its strengthened wing—almost 50% more range than the first-series aircraft.

As the major purchaser of the new C-160s, the French were particularly concerned to ensure that, in line with their commitments outside Europe, all of these aircraft should be capable of in-flight refuelling so as to extend their range for contingency operations. Moreover, they used this opportunity to enhance their overall AAR capability by specifying that 10 of the new aircraft should be equipped as hose and

PLATE 4.11.   C-160 of the Luftwaffe, seen here in Ethiopia in 1985.

drogue tankers, with a further five modified for rapid conversion to the tanker role. Apart from two aircraft used for special communications tasks and another two used for ELINT missions, all of the French Air Force's second series Transalls are primarily employed in the airlift role, crew proficiency in AAR being maintained as a secondary, albeit important skill.

Load-Carrying Capability

With its full-width rear doors, spacious hold and ability to carry cargo up to a maximum weight of 35,275 lb, the fully-pressurised C-160 can carry a wide variety of heavy and bulky payloads, including various combinations of troops and equipment. For example, it can carry up to 93 personnel in the trooping role or, reconfigured for aeromedical duties, up to 63 patients on stretchers with four attendants. In the cargo role, the floor is stressed to carry most types of military vehicle including smaller tanks and light armoured cars, as well as self-propelled guns, missiles and partly disassembled helicopters. A mechanical cargo handling system, with built-in winch and roller conveyor tracks, facilitates the speedy loading and unloading of standard pallets and containers, enabling direct transfer from trucks or other freight-carrying equipment. Loading and unloading can also be assisted by 'kneeling' the main landing gear, thus reducing the slope-angle of the ramp relative to the cargo compartment floor (see Figure 4.3). This system, which makes it easier to move longer

PLATE 4.12.   Pilots' instrument panel on C-160 (second series).

vehicles and loads in and out of the hold, is hydraulically operated through power
supplied either by one of the engines, the APU or by a hand-pump.

The Transall is also highly effective in the air-drop role, being able to deliver 62–
88 paratroops (the precise number depending on their equipment) through large side
doors at the rear of the fuselage. This compares well with the stretched Hercules
which can drop up to 92 paratroops. Alternatively, the C-160 can drop a wide variety
of heavy equipment, including single items up to a maximum weight of 17,637 lb
(8,000 kg). The air-drop techniques employed are described in Chapter 6.

To complete the C-160's all-round tactical versatility, its flight characteristics,
performance and landing gear have been specifically designed for operations from
short, unpaved airstrips. The Transall's ability to undertake such missions, and to
withstand the wear and tear of sustained strip operations in a harsh environment,
was amply demonstrated during the Luftwaffe's participation in the humanitarian

NORMAL   KNEELED DOWN

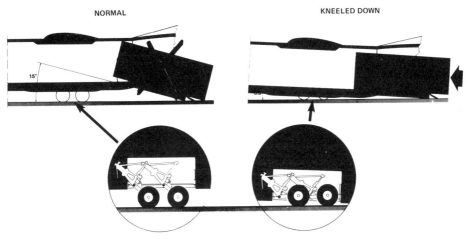

15°

FIG 4.3.   C-160 kneel capability.

PLATE 4.13.   C-160 of French Air Force air-dropping supplies.

airlift of food and medical supplies to remote areas of Ethiopia in 1985. But in addition to its tactical qualities and notwithstanding its primary employment in the short- to medium-range airlift role, the C-160 can fairly claim some degree of inter- as well as intra-theatre capability. Even without the additional fuel tank installed in the second-series aircraft, the C-160 has a range of 2,750 nm (subject to wind component) with a payload of 17,640 lb—quite sufficient for a non-stop crossing of the North Atlantic. While uplift of additional fuel in the extra tank would largely be at the expense of payload (since the maximum all-up weight has not been increased) the provision of an in-flight refuelling capability in all second-series aircraft means that they can accomplish strategic missions with maximum payload, subject only to the availability of tankers.

Summary of Leading Particulars

Flight deck crew: Two pilots, one navigator and one flight engineer. (The second-series aircraft does not carry a navigator.)

External dimensions:	Length	106.3 ft (32.40 m)
	Wingspan	131.3 ft (40.00 m)
	Height	38.3 ft (11.65 m).
Cargo hold dimensions:	Length	44.3 ft (13.51 m)
(excluding ramp)	Maximum width	10.3 ft (3.15 m)
	Maximum height	9.8 ft (2.98 m)
	Volume	4,071 cu ft (115.30 m³).

Engines: Two SNECMA/Rolls-Royce Tyne 20 Mk 22 turboprops, each producing 6,100 eshp and designed for water-methanol injection.
Basic operating weight: 63,934 lb (29,000 kg).
Maximum payload: 35,274 lb (16,000 kg).
Maximum fuel (first series): 32,500 lb (14,742 kg).
Maximum fuel (second series): 47,800 lb (21,682 kg).
Maximum take-off weight: 112,435 lb (51,000 kg).
Maximum landing weight: 103,620 lb (47,000 kg).
Range with maximum payload: 1,000 nm.
Range with 17,640 lb (8,000 kg) payload: 2,750 nm.
Ferry range with maximum fuel (first series): 3,440 nm.
Ferry range with maximum fuel (second series): 4,780 nm.
Cruise speed: 260 kt.
Service ceiling: 26,000 ft.
Typical take-off distance: 2,300 ft (700 m).
Typical landing distance: 2,200 ft (670 m).

F-27 Friendship (Mk400M)

Production of the best-selling Fokker Friendship F-27 recently came to an end after an uninterrupted run of almost 28 years, during which nearly 800 aircraft were built for operators in some 60 countries. Most of these aircraft were commercial transports, the basic F-27 being a short-haul, medium-size, low-cost airliner capable of regular operations into short, semi-prepared airstrips. Such qualities have obvious military applications and, following the early adaptation of commercial models for air force service (see Plate 4.14), a purpose-built military variant—designated the F-27 Mk400M—made its appearance in 1965.

Compared with the version shown in Plate 4.14, which is still in operational service with the Royal Netherlands Air Force 27 years after its acquisition, the Mk400M can fly higher and faster as well as carrying more payload. It also incorporates newer technology and systems, including better avionics. Basically, however, the Mk400M remains true to its pedigree—a well-proven, functional airlifter which is ideal for the

PLATE 4.14.   F-27 of Royal Netherlands Air Force.

smaller logistic support tasks on which it would not be economic to use larger aircraft such as the C-130 and C-160. Unlike most other tactical airlifters, the F27 Mk400M does not have rear doors or a ramp; instead, it has a large cargo and para-door on each side of the rear fuselage, through which all freight must be loaded and from which paratroops and equipment are despatched during air-drop missions. Folding canvas seats along the sidewalls provide space for up to 46 fully equipped troops, whilst in the aeromedical role 24 stretchers can be accommodated in eight tiers of three, plus nine attendants and/or sitting patients. Alternatively, the aircraft can carry up to 13,283 lb of freight or mixed loads of freight and personnel within the available space and weight limits.

Using the normal fuel tanks, the F-27 can carry its full payload of 46 troops over a range of 1,195 nm. If fitted with an additional tank in each wing, which increases the fuel capacity from 9,000 lb to 13,000 lb, the aircraft can carry 28 troops for 1,800 nm, or over a radius of 900 nm without needing to refuel at the destination. Fuel capacity can be increased still further by fitting an external tank under each wing, adding another 3,300 lb to the total fuel carried and thereby increasing the aircraft's maximum ferry range to 2,300 nm.

Summary of Leading Particulars

Flight deck crew: Two pilots, one flight engineer.

External dimensions:

	Length	77.3 ft (23.56 m)
	Wingspan	95.2 ft (29.00 m)
	Height	27.9 ft (8.50 m).

Cabin dimensions:

	Length	47.4 ft (14.46 m)
	Maximum width	8.2 ft (2.49 m)
	Maximum height	6.3 ft (1.93 m)
	Volume	1,978 cu ft (56.0 m³).

Engines: Two Rolls-Royce Dart Mk552 turboprops, each developing 2,210 eshp and

driving a four-bladed constant speed propeller.
Basic operating weight: 25,700 lb (11.655 kg).
Maximum payload: 13,283 lb (6,025 kg).
Fuel capacity (normal): 9,000 lb (4,082 kg).
Fuel capacity (extra wing tanks): 13,000 lb (5,900 kg).
Fuel capacity (extra wing tanks and under-wing tanks): 16,300 lb (7,400 kg).
Maximum take-off weight: 45,000 lb (20,410 kg).
Maximum landing weight: 42,000 lb (19,050 kg).
Range with maximum payload: 1,195 nm.
Ferry range (maximum fuel): 2,300 nm.
Normal cruise speed: 250 kt.
Service ceiling: 27,000 ft.
Typical take-off distance: 2,500 ft (760 m).
Typical landing distance: 2,000 ft (600 m).

G-222

Although the prototype of Aeritalia's G-222 made its maiden flight in 1970, the first production model was not delivered until 1976 (to the Dubai Air Force). This unusually long interval was the result of a protracted trials and evaluation programme entailing several changes in design. However, the final result was an efficient, sturdy, short- to medium-range tactical airlifter which eventually entered service with the Italian Air Force in 1978. With 30 standard transport versions, and another 14 aircraft specifically acquired for the EW, flight calibration and fire-fighting roles, the Italian Air Force is the principal operator of the G-222. Being relatively inexpensive to purchase and maintain, and being particularly well suited to the needs of smaller states in terms of load-carrying capacity and range, the G-222 has been procured by a number of air forces including those of Dubai (part of the United Arab Emirates), Libya, Argentina, Venezuela and Nigeria. Altogether, over 100 aircraft have been delivered and with orders for the transport version now drying up, Aeritalia is believed to be considering the production of Airborne Early Warning, Maritime Patrol/Anti-Submarine Warfare and AAR tanker variants. Meanwhile, the Italian Air Force is also thought to be studying the feasibility of modifying some of its transport versions for an emergency tanker role.

As can be seen from Plate 4.15, the G-222 has the classic features of the contemporary tactical airlifter including a high wing, upswept tail, rear ramp and upward-opening cargo door. Specifically designed for operations from semi-prepared forward airstrips, the landing gear comprises a steerable twin-wheel nose unit with main units each of two single wheels in tandem. The wheels retract into fairings, one of which is visible in Plate 4.15. Although pressure in the oleo-pneumatic shock absorbers can be adjusted to vary the height and attitude of the cargo floor in relation to the ground, unloading of some vehicles can still be difficult. This is well illustrated in Plate 4.16 which shows an aerial port team trying to deal with the problem of ground clearance during the unloading of an ambulance from a G-222 during a recent NATO exercise in Turkey.

In the passenger role, the aircraft (which is fully pressurised) can carry a maximum of 53 fully equipped troops—32 in foldaway sidewall seats plus a further 21 in

PLATE 4.15.   G-222 of Italian Air Force at Erzurum, Turkey, June 1987.

PLATE 4.16.   Aerial Port Team having problems in unloading an ambulance which
has run out of ground clearance leaving the aircraft ramp.

stowable seats. The Italian Air Force has a number of quick-change kits which enable
the main cabin to be readily converted into the aeromedical role, a configuration in
which 36 stretchers and four medical attendants can be carried. The size of the cargo
hold, coupled with the aircraft's weight-carrying capability, permit the airlift of

various permutations of troops and equipment, including:

- 2 × CL-52 light trucks.
- 1 × CL-52 light truck with 105-mm howitzer.
- 1 × AR-59 recce vehicle with 106-mm gun.
- 5 × standard pallets.
- 1 × ambulance plus 16 troops.

Although primarily employed in the airlanding role, the G-222 also has a useful air-drop capability. The rear doors can be opened in flight to permit the despatch of up to five pallets each weighing up to 2,205 lb (1,000 kg) or a single pallet weighing up to 11,023 lb (5,000 kg). Up to 40 fully equipped paratroops (32 on sidewall seats plus eight on stowable seats) can be dropped, either over the rear ramp or through side doors towards the rear of the fuselage.

The standard G-222 has a comprehensive and modern avionics suite which includes an autopilot, flight director system, radar altimeter, twin VOR, twin TACAN, twin ILS, as well as OMEGA and a weather/ground mapping radar. Combined with the aircraft's de-icing system, this equipment theoretically confers an all-weather capability but this is not always the case in practice since the G-222's relatively low cruising altitudes (typically 15,000–18,000 ft depending on weight and temperature) mean that it cannot always avoid the larger storm cells. When severe and extensive weather systems are forecast to lie across the intended track, G-222 missions may have to be diverted, postponed or aborted.

Summary of Leading Particulars

Flight deck crew: Two pilots and one radio operator/flight engineer.

External dimensions:	Length	74.5 ft (22.70 m)
	Wingspan	94.2 ft (28.70 m)
	Height	32.2 ft (9.80 m).
Cargo hold dimensions:	Length	28.2 ft (8.58 m)
	Width	8.0 ft (2.45 m)
	Height	7.4. ft (2.25 m)
	Volume	2,048 cu ft (58.0 m³).

Engines: Two (Fiat-built) General Electric T64-GE-P4D turboprops, each rated at 3,400 eshp at ISA + 25°C and driving three-bladed variable-pitch, reversible propeller.
Basic operating weight: 34,610 lb (15,700 kg).
Maximum payload: 19,840 lb (9,000 kg).
Maximum fuel: 20,725 lb (9,400 kg).
Maximum take-off weight: 61,730 lb (28,000 kg).
Maximum landing weight: 58,420 lb (26,500 kg).
Range with maximum payload: 740 nm.
Range in aeromedical role (36 stretcher patients plus four attendants): 1,340 nm.
Ferry range (maximum fuel): 2,500 nm.
Normal cruise speed: 195 kt.
Service ceiling: 25,000 ft.
Typical take-off distance: 2,460 ft (750 m).
Typical landing distance: 1,800 ft (550 m).

C-23A Sherpa

Shorts of Belfast has been producing small, functional airlifters for both military and civilian operators for over 20 years. One of its best-known products is the Skyvan, developed in the 1960s. A compact, sturdy aircraft, the military version of the Skyvan (designated Srs3M) has proved to be ideal for airlifting small quantities of supplies or troops over short distances using rough strips which are too small for larger aircraft. In the mid-1970s, Shorts produced a larger and more powerful airliner (designated 330-200). Derived from the Skyvan, and aimed at the smaller 'commuter' or 'feeder' airlines, the 330-200 is a twin turboprop designed to carry 30 passengers on short-haul flights into and out of semi-prepared strips. The Sherpa, which first flew in 1984, is the freighter version of the 330-200, the only significant differences being the addition of a rear-loading ramp and door, and the installation of a strengthened floor. Eighteen military variants of the Sherpa (designated C-23A) were delivered to the USAF in 1984/85.

Although tactical transport operations are predominantly geared to the deployment and support of ground forces, this is not always the case, the C-23A affording an excellent illustration of how tactical airlift can also be used to provide essential logistic support for air combat units. This is the role for which the C-23A has been specifically acquired by the USAF, its mission being to ferry critical and high priority spares, including engines, between US fighter bases and depots in Europe in both peace and war. By mounting a combination of scheduled and quick-response special flights—and, in war, operating as necessary from grass areas, taxiways or surviving

PLATE 4.17.   USAF C-23A Sherpa.

sections of the runways at bomb-damaged airfields—the C-23A's employment on this vital task ensures that US fighter strength in Europe can be maintained at its optimum level. This is a good example of how tactical airlift can be used as a cost-effective force multiplier of air power as a whole.

As can be seen from Plate 4.18, the cargo compartment is functional and uncluttered. Freight can be trans-loaded through either the large forward door or full-width, hydraulically operated rear ramp/door (see Plates 4.19 and 4.20). The ramp/door, which is self-supporting and stressed to carry 5,000 lb, can be adjusted through a range of positions to simplify loading from different types of freight handling equipment. A roller conveyor system, used to facilitate the loading of pallets and containers, can be quickly removed to permit the installation of up to 18 passenger seats. Otherwise, the cargo compartment contains only six foldaway seats. Typical loads could include three containers and five passengers, two Land-Rover type vehicles or an F-111 engine. In view of its all-weather mission and the high-intensity air traffic environment of its theatre of operations, the C-23A has a more sophisticated avionics suite than most short-range transports. Its comprehensive equipment includes UHF, HF and dual VHF AM/FM radios; twin flight directors; autopilot; twin ADF; twin VOR/ILS; LTN-96 INS; TACAN; IFF transponder; Collins WXR-300 colour weather radar with terrain mapping; ground proximity warning system; and radar altimeter.

PLATE 4.18.   C-23A cargo compartment.

PLATE 4.19.   Forward door of C-23A.

PLATE 4.20.   Loading in progress through rear door of C-23A.

Summary of Leading Particulars

Crew: Two pilots and one flight mechanic/airloadmaster. (The airloadmaster's duties include routine servicing at *en route* stops as well as overseeing the loading and off-loading of cargo.)

External dimensions:

Length	58.0 ft (17.68 m)	
Wingspan	74.7 ft (22.77 m)	
Height	16.3 ft (4.97 m).	

Cargo compartment: Length   29.8 ft (9.09 m)
                               Width   6.5 ft (1.98 m)
                               Height   6.5 ft (1.98 m)
                               Volume   1,260 cu ft (35.68 m³).

Engines: Two Pratt & Whitney PT6A-45R turboprops, each producing 1,198 eshp and driving a five-bladed, constant-speed propeller.
Basic operating weight: 14,727 lb (6,680 kg).
Maximum payload: 7,000 lb (3,175 kg).
Fuel capacity: 4,480 lb (2,032 kg).
Maximum take-off weight: 22,900 lb (10,387 kg).
Maximum landing weight: 22,600 lb (10,251 kg).
Radius of action with maximum payload: 195 nm.
Range with 5,000 lb payload: 660 nm.
Normal cruise speed: 157 kt.
Typical take-off distance: 3,400 ft* (1,030 m).
Typical landing distance: 3,000 ft* (900 m).

*with maximum payload.

Questions

1. Why do most states with armed forces maintain at least some tactical transport capability?

2. List five important qualities required of a tactical airlifter.

3. What are the main design features of contemporary tactical airlifters?

4. (a) Why are aircraft such as the C-130 particularly suitable for adaptation to non-transport missions?

   (b) List five non-transport roles on which the C-130 has been employed.

5. What are the main operational limitations of the AN-12 Cub?

6. What features distinguish the second-series C-160 from the earlier model?

7. (a) Which of the tactical airlifters described in this Chapter does not have rear doors to facilitate loading and unloading?

   (b) Which of the tactical airlifters described in this Chapter have a 'kneel' capability?

8. Give an example of how tactical airlift can be used as a 'force multiplier' of air combat units.

5

Factors Involved in Planning Airlift Operations

To the uninitiated, the planning and execution of military air transport operations may seem a relatively undemanding process. After all, what—on the face of it—could be more straightforward than mounting a mission or series of missions from A to B with none of the 'applied' flying, involving sensors or weapons systems, inherent in most other air operations? Any such perceptions are, of course, both simplistic and illusory. All transport missions, whether strategic or tactical, scheduled or non-scheduled, routine or complicated, are subject to a number of special factors which vary with the nature of the particular task in question. It is not only the airlift planner and immediate operating authority who must understand and take account of these factors; strategic planners and military commanders at the highest levels should also be aware of their existence and implications if air transport assets are to be exploited to best effect. In the wider context, no serious student of air power who wishes to develop a sound grasp of his subject can afford to disregard the special parameters which condition the planning of airlift operations.

NEED FOR JOINT PLANNING

It is most important to remember that airlift operations are not mounted for their own sake but always in support of other air, land or seaborne forces. Even though it may not always be possible to meet the user's requirements in full, they must be fully understood and if necessary clarified at the outset of the planning process. It is not sufficient for airlift planners merely to know outline parameters such as the airfields of departure and destination and the type and amount of payload to be carried; they must also be aware of the specific aims of the wider operation to which the airlift relates, as well as the broader politico-military context. Above all, planning must be closely coordinated with all of the interested parties from the very beginning. Otherwise, those providing the airlift may misunderstand the objectives of the customers, while the customers may fail to appreciate the factors and constraints which must be considered in the production of a mutually acceptable and workable plan. The more significant of these factors and restrictions are examined below.

POLITICO-MILITARY CONTEXT

As noted in Chapter 1, airlift potential can be exploited to exert political as well as military influence in certain circumstances. Conversely, air transport operations must themselves take careful account of the prevailing politico-military situation, especially if they are to be conducted beyond the national borders of the participating state or states, and against a background of heightened or rising tension. Even in conditions loosely described as 'peacetime', planners must consider this wider context. Although there have been no wars in Europe since 1945, many have erupted elsewhere, some strictly limited in duration and effect, but others—such as the seven-year conflict between Iran and Iraq—threatening to drag on indefinitely and inconclusively with serious consequences both for neighbouring states and for the international community. Not only could exercises or operations in the vicinity of such areas be interpreted or characterised as provocative by one or more of the protagonists; they could also prove hazardous to any transport aircraft involved.

AIR SITUATION

By virtue of their size, performance, relative unmanoeuvrability and lack of armament (with the exception of Soviet aircraft such as the AN-12 and AN-22 which are sometimes equipped with a rear turret containing twin cannon) all transport aircraft are highly vulnerable to hostile action. Their degree of susceptibility depends on their mission profile and the extent to which they are actually threatened by enemy air or ground forces. In some scenarios, even where an enemy enjoys air superiority or a

PLATE 5.1.   Workhorse of the USAF's MAC, the Lockheed C-141B Starlifter is able to extend its already considerable range by in-flight refuelling, a facility which allows it to avoid sensitive airspace or, if necessary, to operate along routes which do not require diplomatic clearance.

favourable air situation, there could conceivably be circumstances in which the risks of an air transport operation are outweighed by the likely benefits of a successful outcome. For example, operational imperatives might justify the night-time insertion of a special forces unit by parachute, using clandestine techniques. In general, however, planners should avoid committing transport aircraft to operations in areas where friendly forces do not enjoy at least some semblance of a favourable air situation, if only for limited periods. This constraint applies not merely to destination airfields or dropping zones but equally to the *en route* phase of an airlift operation. Thus, in 1987, the planners of an exercise deployment of troops to Egypt from Western Europe would have been prudent to select a route well clear of Libyan airspace in order to minimise the opportunity for unpredictable factions within that country to threaten or worse still attack the transport aircraft concerned. This is but one example of many in the modern international climate. In other parts of the world, political sensitivities or the latent hostility of some states towards others, place similar restrictions on the safe movement of military transport aircraft. Irritating though such constraints might be—entailing, for example, expensive and time-consuming deviations from the most direct route—they cannot be ignored. If a transport aircraft is unnecessarily exposed to hostile action and subsequently lost, the customer not only loses that particular payload of personnel and equipment but also all further usage of that aircraft for future missions.

TYPE OF MISSION

Having taken due note of the latest intelligence reports (which will indicate the prevailing political framework and likely air situation) the planners must next address the following series of questions designed to establish the basic operational parameters within which the airlift is to take place:

- What is the aim of the exercise or operation?
- Does the exercise or operation represent the execution of an existing contingency plan?
- When will the exercise or operation commence?
- What is the expected duration?
- What is the timescale within which the airlift must be completed?
- What is the nature and size of the airlift task?
- What number and type of missions does this entail?
- Does the task involve the deployment of air combat units and, if so, do they require AAR?
- How does the airlift requirement relate to any requirement for surface movement?
- How does the combined air/surface movement plan relate to the overall operational plan?
- What are the preferred departure and destination airfields?
- Are the destination airfields (or dropping zones) secure?
- Is the airlift mission to be mounted overtly or is it to be unobtrusive, perhaps with a view to achieving surprise?
- Will the force require subsequent resupply by air and, if so, to what extent?

It is essential to obtain early and unambiguous answers to these questions in order to determine whether the proposed airlift mission is feasible, to assess broadly the resources needed for its successful completion, and to provide a basis for the detailed further planning that must follow. The timeframe is particularly important. If, for example, the operation is to be mounted at short notice, airlift assets may need to be rapidly recalled from other tasks and assembled at the departure airfield. Even where the execution of an existing contingency plan is involved, this will nevertheless have to be carefully checked. Few if any airlift contingency plans can be implemented as an off-the-shelf package, however carefully they may have been formulated and updated. Almost invariably, such plans will need to be modified to take account of actual circumstances and operational factors that may differ significantly from those envisaged when the plan was drawn up.

AIRFIELD REQUIREMENTS

Before an airlift can be mounted, it is essential to establish whether the proposed staging and reception airfields are capable of handling the intended flow of transport aircraft. Drawing upon various sources—which could include unclassified aeronautical directories, intelligence material, or actual surveys of the airfields in question—the planner needs answers to two fundamental questions:

- Which types of aircraft can the airfield physically accept?
- What infrastructure and facilities are available to support an airlift operation?

In order to establish these facts, a great deal of highly detailed information is required. For example, the type and number of aircraft that can be used will depend, *inter alia*, on the dimensions, condition and load-bearing strength of the runways, taxiways and parking ramps. The extract reproduced in Figure 5.1 from a comprehensive checklist used by some air forces to conduct airfield surveys illustrates the detailed categories of information that must, if possible, be obtained by the planners.

AIRFIELD STRENGTH

It is not enough to confirm that the dimensions of an airfield's runways, taxiways and parking areas are adequate; it is also important to ensure that the load-bearing strength of these various surfaces is compatible with the weight 'footprint' of aircraft earmarked for operations at that particular airfield. Repeated or even infrequent operations by heavy aircraft, such as C-141B Starlifters, on an airfield surface which has not been constructed to withstand such weights, could cause serious damage both to the airfield and to the aircraft itself. Several methods are used to ensure that maximum weights are not exceeded, the most common being the Load Classification Number (LCN) and Load Classification Group (LCG) system, which is recognised by the International Civil Aviation Organisation (ICAO) and widely used throughout the world. This system establishes a safe relationship between the strength of a paved surface, based on its construction qualities and expressed in terms of LCG, and the weight of an aircraft expressed in terms of LCN. As the name suggests, an LCG indicates the range of LCNs which a particular surface can normally accept on a

---

Runway data

a. Number of runways.
b. Runway directions (magnetic bearings).
c. List following for each runway.

    (1)   Length (ft).
    (2)   Width (ft).
    (3)   Type of surface.
    (4)   Condition.
    (5)   Slope.
    (6)   Published strength (LCG).
    (7)   Any restrictions on published strength.
    (8)   Centreline markers?
    (9)   Distance markers?
    (10) Approach lighting?
    (11) Threshold lighting?
    (12) Runway lighting?
    (13) Visual Approach Slope Indicator System (VASIS)?*
    (14) Shoulders – Type of surface.
                       Width (ft).
                       Condition.
    (15) Overrun – Length (ft).
                  Type of surface.
                  Condition.
    (16) Any significant obstructions adjacent to runway ? If so, describe position, type and height in feet.

*A system of lights, positioned each side of the runway near the touch-down area, which provides an accurate visual indication of the aircraft's vertical position on the glidepath.

FIG. 5.1   Example of Runway Checklist used for Airfield Survey

regular basis. LCNs for individual aircraft types and LCGs for specific airfields are published in the relevant aeronautical documents.

Having noted the LCNs of the aircraft he proposes to employ in a particular operation, the planner must check (using the table reproduced as Table 5.1) that these fall within the range of LCN values covered by the published LCGs of the airfields involved. If so, the aircraft may make unlimited use of the pavements at those airfields. However, if its LCN falls within a bracket one higher than the published LCG, an aircraft may only use that airfield occasionally and then strictly

TABLE 5.1
LCN/LCG Table

Aircraft	Airfield
LCN	LCG
101–120	I
76–100	II
51–75	III
31–50	IV
16–30	V
11–15	VI
0–10	VII

Examples
An aircraft with an LCN of 70 may operate on an unrestricted basis from airfields of LCG III, II or I. However, such an aircraft may only operate from airfields of LCG IV on an occasional basis, and from airfields of LCG V only in an emergency. It should not operate into airfields of LCG VI or VII at all.

subject to the approval of the airfield authorities. Aircraft with an LCN which places it two categories higher than an airfield's published LCG may only operate into such an airfield in an emergency. These rules are illustrated in the examples given in Table 5.1.

Although this is a well-tried system, it does rely on the availability of accurate airfield data. Unfortunately, the strength characteristics of some airfields, especially those in less developed parts of the world, are either inaccurate, arbitrarily established or simply not known. The planner who wishes to use such airfields may therefore have to estimate their weight-bearing capacity, basing his judgement if possible on empirical evidence of previous usage by aircraft with LCNs similar to those of the airlifters in question.

OTHER OPERATIONAL FACTORS

Navigation Aids

In addition to confirming the size and load-bearing capacity of staging and destination airfields, the planner must investigate a number of other important operational factors, the most crucial of which is the availability of navigation aids and facilities. It is axiomatic, yet sometimes overlooked, that all other considerations are academic unless transport aircraft can approach, land and depart expeditiously and safely in all conditions likely to be encountered during a particular operation. The specific aids required will depend, of course, on the airfield's location, expected meteorological conditions, time of year and type of operation (eg, all-weather or day/Visual Flight Rules (VFR) only). In some cases, where local terrain and weather are especially favourable, and where the pattern of missions is not unduly intensive, it may be possible to mount an airlift with no navigation aids other than those carried in the aircraft, provided that adequate air traffic control is available to ensure the necessary procedural clearances and separations. More often than not, however, the planner will need to confirm the availability of at least one 'approach aid' and/or 'landing aid':

VHF Omnidirectional Radio Range (VOR) Beacon

VOR beacons are employed all over the world. Signals continuously transmitted from a ground installation are interpreted by a special receiver in the aircraft where the information is displayed on the pilot's instrument panel to indicate the aircraft's bearing (or radial) from the beacon. Although used mainly for short-range navigation along airways, VOR beacons positioned on or near airfields may also be used as an approach aid. Having fixed his position overhead the VOR, the pilot can then use a combination of bearing information and timing to descend safely (if necessary in adverse weather) to a predetermined height and position at which he can either continue the approach and landing using visual cues or, if this is impossible, initiate the relevant 'missed approach' procedure.

Tactical Air Navigation (TACAN) Beacon

While not as common as VOR systems, TACAN beacons are also widely used as an aid to navigation, especially for military purposes. In simple terms, the aircraft interrogates a transponder-type beacon on the ground; the beacon then transmits UHF signals which, after processing by special on-board equipment, enable the crew to determine their position in terms of distance and bearing from the ground installation. While normally employed for short-range navigation, TACAN beacons on or adjacent to airfields can also be used in much the same way as VOR beacons to carry out a safe descent and approach to a predetermined height and position on the final approach to land.

VORTAC Beacon

This is simply a navigation aid in which VOR and TACAN beacons are co-located at the same ground position. The VOR installation provides azimuth (or bearing) information, while distance is derived from the 'distance measuring equipment' (DME) element of the TACAN array. As with separate VOR and TACAN beacons, suitably positioned VORTACs can be used as an airfield approach aid.

Instrument Landing System (ILS)

The ILS is a pilot-interpreted landing aid which allows an aircraft to be flown with a high degree of accuracy, in all weathers, on the final approach. The main ground-based installation produces two radio beams: one is a 'localiser' beam which provides azimuthal (or lateral) guidance relative to the centreline, and the other is a 'glideslope' beam which provides guidance relative to the glidepath (normally angled at $3°$ to the horizontal). Most ILS systems also incorporate 'outer' and 'inner' marker beacons, positioned on the ground under the datum approach path, which trigger audio-visual cues on the pilot's instrument panel to confirm the aircraft's position and distance to touch-down. Collectively, the ILS beams and marker beacons enable the pilot to fly his aircraft down to a predetermined and relatively low height and position on the final approach, from which he can either complete the landing using visual cues or, if this is impossible, initiate an overshoot and 'missed approach'.

Precision Approach Radar (PAR)

PAR is a primary radar system used to control aircraft from the ground during the initial and final approach phases of landing. By determining the aircraft's range from touch-down, and its lateral and vertical displacement from the datum approach path, PAR enables the ground controller and pilot jointly to achieve a highly accurate approach and landing, even in adverse weather. The procedure is very straightforward. Having established radio contact with the pilot and positively identified his aircraft, the PAR controller uses the information displayed on his radar screen to direct the aircraft onto the centreline and glidepath to a predetermined and relatively low height and position close to the runway threshold from which the landing can either be completed using visual cues or, if this is impossible, the pilot must initiate an overshoot and carry out the relevant 'missed approach' procedure.

Unless an airfield is equipped with one or more of the above aids, it may prove impossible to maintain the requisite flow of aircraft by day and night in all weather. Hence, if an airfield is not already equipped to the necessary standard, air-portable navigation aids may have to be flown in before the main airlift can begin. Naturally, this increases the overall airlift bill; for example, three C-130s are required to position a typical air-portable PAR.

Emergency Facilities

If justified by the operational situation, transport aircraft would operate into forward airfields whether or not the airfields had adequate fire and crash facilities to deal with local emergencies. In peacetime, however, transport operations would not normally be mounted into airfields without the necessary fire and medical cover. In general, larger aircraft such as the C-5 Galaxy and C-141B Starlifter require more fire and medical cover than smaller aircraft such as the C-130 Hercules. This is because the sheer size of the bigger airlifters and their greater passenger-carrying potential would make an on-board fire potentially much more serious and difficult to extinguish. The planner must therefore ensure not only that emergency services are available at staging and destination airfields, but also that the type and numbers of fire-fighting vehicles and ambulances (and crews to man them around the clock if necessary) are appropriate to the aircraft types involved. If the required emergency services are not already available, they too may have to be pre-positioned ahead of the main airlift.

Ground Manoeuvrability

Once satisfied that aircraft can operate safely from the airfields he intends to use, the planner must consider certain aspects of their manoeuvrability on the ground. For example, it is important to know—especially if the main parking ramp is likely to be congested, or to have to cope with a considerable volume of traffic—whether parking spots are marked out and if remote spaces are available for the unloading of ammunition and explosives. For safety reasons, such cargo should always be off-loaded in designated areas well away from the main parking ramp. Where aircraft are to be parked on a slope, it is helpful to know in advance whether this is likely to increase the engine power needed to achieve 'break-away' from the static position and to maintain taxi momentum. If so, it is necessary to ascertain whether engines can be run up to maximum power on the ramp without damaging nearby structures or dislodging loose surfaces to an unacceptable extent. Similarly, it is prudent to check whether the application of take-off power at the runway thresholds is likely to cause slipstream damage to adjacent structures, surfaces or vehicular traffic.

SUPPORT FACILITIES

Although operational considerations are overriding, the ability of a given airfield to sustain an airlift operation also depends heavily upon its support facilities and infrastructure.

Aerial Port

When aircraft have reached their destination, the first priority is to off-load their passengers and/or cargo so that these can be deployed into the forward operational areas as rapidly as possible. In NATO, the task of unloading and reloading aircraft is usually undertaken by a team of specialist personnel who collectively form what is known as the 'aerial port' element. Troops are relatively easy to process, especially if they can deplane via integral aircraft ramps (if fitted), thus obviating the need for passenger steps. The off-loading of heavy cargo, however, requires specialist expertise and ground handling equipment including fork lift trucks and low-loaders (see Plates 5.2 and 5.3). If aerial port personnel and equipment are not already available at reception airfields, they will have to be pre-positioned before the airlift proper can begin.

Maintenance

In principle, destination and staging airfields are normally expected to offer only basic maintenance facilities for transport aircraft, such as marshalling teams, and personnel to undertake after-/pre-flight inspections and minor servicing. The planner must ensure that this level of technical support, with the requisite equipment, is available and that the range of expertise covers all types of aircraft involved in the airlift operation. It is often impossible to satisfy this requirement from existing in-place resources, in which case a specialist maintenance team will need to be deployed to the airfields involved for the duration of the operation. In the event of a major unserviceability (requiring, for example, an engine or propeller change) the necessary

PLATE 5.2.   Heavy-duty cargo loader being used to move a Luftwaffe pallet.

PLATE 5.3. Typical fork-lift cargo loader, seen here being operated by a member of
a USAF Aerial Port Unit.

spares, equipment and personnel would have to be flown in from the aircraft's home
base. Such contingencies do happen (see Plates 5.4 and 5.5). All that the planner can
do in advance is to recognise this as a fact of life and identify—during his initial
appraisal of airfield resources—any hangars or other maintenance facilities that
might be available if needed.

Refuelling and De-icing

In order to keep turn-round times to a minimum and to reduce the logistic support
needed in the forward area, aircraft engaged in airlift operations are not normally
refuelled at destination airfields unless the flow of missions is very light and some
refuelling is essential to enable aircraft to recover safely to another base. In cases
where aircraft cannot carry sufficient fuel from the point of departure for the round
trip, the planner must endeavour to route them through staging airfields where they
can refuel outbound and/or inbound as required and, if necessary, change or 'slip'
crews. If the non-availability of staging posts necessitates refuelling at the destination,
the amount uplifted should be as small as possible. At the planning stage, it is
nevertheless important to identify what refuelling facilities, if any, are available. In
particular, it is essential to know what grades and quantities of fuel, and what types
and numbers of refuelling points and vehicles, are available. Similarly, if cold-weather
operations are anticipated, the planner must ensure that adequate snow clearance
and de-icing facilities can be provided.

PLATES 5.4 and 5.5. Engine change in progress on Belgian Air Force B-727 at Erzurum, Turkey.

Communications

As with most other military activities, the successful execution of air transport operations depends heavily on effective communications between the key agencies involved at all levels of the command and control organisation. Specifically, the airlift planner must ensure that there are appropriate links between the Airlift Coordination Centre (ALCC)—or such other agency as may be set up to coordinate the airlift operation—and the staging and reception airfields; between those airfields themselves; and between the ALCC and the national air transport force headquarters.[1] Satellite communication (SATCOM) links are particularly effective for this purpose. Compact, portable and simple to set up and operate, SATCOM offers long-range, secure, data and clear voice links between all units equipped with the necessary terminals. Both the control element at the reception airfield and the transport force headquarters must also have 'flight watch' HF radio links with participating aircraft, in order to monitor and where necessary adjust the flow. At the destination airfield itself, there must be effective links between the operations, air traffic control, aerial port, maintenance and meteorological elements, and between the overall control element and the force reception authorities. A combination of telephones, land-lines and hand-held radio sets is usually employed for this purpose.

PLATE 5.6. De-icing in progress on C-141B at Bardufoss Airfield, northern Norway, 1988.

PERFORMANCE PLANNING

Except where operational circumstances dictate otherwise, airlift missions must always be planned and executed in such a way as to ensure maximum safety of the aircraft involved. Hence, in addition to confirming the various airfield requirements, the airlift planner must also consider whether aircraft performance, in terms of maximum gross weight and appropriate safety margins, is sufficient for successful completion of the projected task. The basic objective of performance planning (which is an essential feature of all modern air transport operations) is to ensure in advance of any flight that the room needed to manoeuvre an aircraft safely is never greater than the space available in the event of:

- Incident-free flight, or
- An incident such as the failure of a power unit on take-off, landing, or whilst *en route*.

In essence, performance planning seeks to ensure that the room required for a manoeuvre is never more than the space available. Since the room required depends upon and increases with weight—the primary controllable variable in the equation— the principle net product of a performance plan is the maximum permissible take-off weight. This in turn determines the payload that can be carried on each mission after making due provision for fuel.

For performance planning purposes, flights are divided into four stages, defined as follows:

- *Take-Off*: From the beginning of the take-off run to an initial height of 35 ft (50 ft for some aircraft).
- *Take-Off Net Flight Path*: From a height of 35 (or 50) ft to a height of 1,500 ft above the airfield.
- *En Route Stage*: From a height of 1,500 ft above the departure airfield to a height of 1,500 ft above the destination airfield.
- *Landing*: From a height of 1,500 ft above the destination airfield to the point on the runway where the aircraft completes its landing run.

Data is derived from a statistical analysis of test flights made to determine gross performance for each of these four stages under various conditions of aircraft weight, airfield altitude and temperature, runway slope and wind component. Gross performance is then suitably factored to allow for various contingencies (such as unavoidable variations in piloting technique) to produce a minimum standard termed 'net' performance. All planning is based on net data except when military operating standards are applied.[2]

Airlift planners must calculate the heaviest permissible take-off weight from the net performance data after considering:

- The aircraft's structural limitations (imposed by the manufacturer).
- Weight, altitude and temperature (WAT) limits for take-off and landing.
- Airfield criteria (eg, runway lengths and gradients).
- Take-off net flight path.
- *En route* terrain clearance.
- Landing weight at destination.

Additional constraints on take-off or landing weights (related to aircraft type or imposed by the airfield authorities) can include:

- Brake heating limitations.
- Tyre speed and pressure limitations.
- Cross-wind limitations.
- Use of 'bleed air'[3] for de-icing or air-conditioning systems.
- Reduced performance due to slush, snow or standing water on the runway.
- Airfield pavement strength.
- Aircraft anti-skid system inoperative.
- Reverse thrust inoperative.
- Flap settings.
- Noise abatement regulations.

TAKE-OFF PERFORMANCE

Take-off performance is calculated on the basis that, should one power unit fail before the take-off stage of flight has been completed, there is a sufficient margin to continue the take-off on the remaining engine or engines. This pessimistic but prudent assumption ensures that, in the unlikely event of an actual failure during this critical phase of flight, the take-off can be safely completed provided that the failure occurs at or above the aircraft's pre-calculated 'decision speed' (V1). The V1 speed, which depends on aircraft weight and airfield dimensions, is calculated before every flight. In the event of single engine failure, V1 is the speed below which take-off must be abandoned, and above which take-off must be continued. Hence reference to V1 enables the pilot to make a purely objective decision whether to continue or abort if an engine failure or some other emergency occurs during the take-off run.

The maximum permissible take-off weight for any flight is the lowest weight calculated after considering the following:

Certificate of Airworthiness Limits

Maximum gross weights for take-off and landing are specified in each aircraft type's 'Certificate of Airworthiness', issued by the national authority in the country where the aircraft is registered. Governed by structural considerations, these figures are absolute. The take-off limit should never be exceeded, while the landing limit should only be exceeded in emergency. Although it must always be taken into account, the limit on landing weight seldom restricts take-off weight in practice, in view of the reduction in aircraft weight which occurs as fuel is burned during flight.

Take-Off Weight, Altitude and Temperature (WAT) Limit

In general, aircraft performance declines with increases in ambient temperature and airfield altitude. In practical terms, this means that most transport aircraft can operate at significantly higher weights from 'low and cool' airfields such as London and Frankfurt than from 'hot and high' airfields such as Mexico City (7,340 ft above mean sea-level, with an average daily maximum temperature of $+26°C$) and Nairobi

(5,308 ft above mean sea-level, with an average daily maximum temperature of +25°C). In order to carry as much payload as possible from 'hot and high' airfields such as these, an aircraft might need to take off when ambient temperatures are at their lowest (probably just before dawn) and/or depart with minimum fuel with a view to topping up tanks at an intermediate (and ideally a lower and cooler) airfield *en route*. The 'WAT Limit' is effectively the maximum take-off weight for a particular combination of airfield altitude and temperature. Compliance with this limit—which can be easily calculated from the relevant operating data manual—ensures that an aircraft can achieve specified gradients of climb during the take-off net flight path stage with one power unit inoperative.

Take-Off Distance Requirements

Both planner and pilot must ensure that there is adequate runway length and clearance from obstructions to permit a safe take-off with one engine inoperative, taking due account of aircraft weight, airfield altitude and temperature, runway slope and condition (ie, wet or dry) and wind component. The actual procedures involved in determining field length requirements can be rather complicated, although some air forces have simplified the process by devising easy-to-use graphs and tables or by developing computer programmes for use during pre-flight planning.

Take-Off Net Flight Path Performance

The 'Take-Off Net Flight Path' is the profile which an aircraft is expected to achieve during its climb-out to a height of 1,500 ft with one power unit inoperative. In most countries, it is a legal requirement that this profile should ensure a vertical clearance of at least 35 ft above all obstacles which fall within a specified area (known as the 'obstacle domain') either side of the aircraft's intended track. The required vertical interval increases to 50 ft during any lateral turn of more than 15° during the departure procedure. In order to satisfy himself that he can comply with this important safety regulation, the pilot uses special 'obstacle clearance charts' during flight planning to ascertain whether his aircraft's performance during the Take-Off Net Flight Path stage (which depends on aircraft weight, airfield altitude and temperature, and runway slope and wind component) is sufficient to achieve the necessary clearances. If the calculations show that these clearances cannot be achieved, then either the aircraft weight will have to be reduced or the pilot may be able to plan his departure using another runway and/or climb-out procedure where the obstacle criteria are less critical. In essence, the purpose of calculating Take-Off Net Flight Path performance is to establish the maximum weight at which, in the event of a single engine failure, the aircraft can safely climb to 1,500 ft avoiding all obstacles by the prescribed clearances.

En Route Terrain Clearance

Most countries stipulate that transport aircraft, military as well as civilian, must be able in the event of single engine failure to maintain certain minimum vertical clearances above obstacles along their intended track. For example, British air navigation regulations require transport aircraft with one engine inoperative to clear all obstacles within 10 nm of track (or 5 nm, if the aircraft is able to maintain track to this degree of accuracy) by at least 2,000 ft. Ability to comply with this regulation depends mainly on aircraft weight. Hence the planner must calculate whether the height of the terrain or other obstructions along the intended track, balanced against the expected reduction in weight as fuel is burned off, and performance at the anticipated weights and temperatures, is such as to impose any restriction on take-off weight.

Landing Weight, Altitude and Temperature (WAT) Limit

Just as there is a WAT limit for take-off, so there is a WAT limit for landing to ensure that the actual aircraft weight during the final stage of flight does not exceed the maximum landing weight governed by the destination airfield's altitude and forecast ambient temperature at ETA. This limit ensures that, with one engine inoperative during the landing phase, the aircraft will be able to maintain the required gradient of climb in the event of a 'missed approach'. The landing WAT limit must be calculated before departure but, in view of the fuel burn and consequent reduction in weight during the flight, is unlikely to restrict the take-off weight.

Landing Distance Requirements

The basic aim in calculating these requirements is to ensure that—for a given combination of aircraft weight, airfield altitude and temperature, runway slope and wind component—the landing distance required does not exceed the landing distance available at either the destination or alternate airfield.

Return Sector Requirements

Since the destination airfield will itself become a departure airfield once passengers and cargo have been offloaded, all the factors outlined above must also be considered in respect of missions returning from the destination. This is particularly important when planning to use airfields at which aircraft can land with a substantial payload, but not take off at the high weights which might be necessary in order to carry a similar payload on the return sector. For example, an RAF C-130 could carry a heavy payload from Akrotiri in Cyprus to Nairobi, but would probably be unable to airlift a similar payload back to Akrotiri without refuelling at an intermediate airfield *en route*.

MISSION PLANNING FACTORS

Having established that the task is feasible in terms of airfield facilities and aircraft performance, the next step is to consider a number of practical factors which will determine the final concept and shape of the airlift plan.

Selection of Route

As already noted in the section on the prevailing air situation, determination of the route is all-important. The fundamental aim is to select a route which will allow flights to be completed in the minimum time, thereby completing the overall mission as rapidly as possible whilst economising in valuable flying hours and fuel. The actual route selected may be influenced by one or more of the following.

Overflight Clearance

Military transport aircraft are not normally allowed to operate through foreign airspace without specific overflight clearance from the state or states concerned. If the required approval—usually referred to as 'diplomatic clearance'—is denied, the planner may have to accept significant deviations from his preferred route, increasing the duration and cost of the mission and perhaps necessitating the use of AAR or staging airfields. A number of the world's air forces have had to contend with such problems in recent years. For example, the USSR has sometimes had to route military transports around Norway's North Cape *en route* to Cuba, while the USA has sometimes been denied overflying rights, when mounting resupply flights to Israel, by countries sympathetic to the Arab cause.

Airspace and Air Traffic Control

In war, air traffic control would probably be placed under military direction in order to coordinate the complex and competing airspace requirements of air defence systems and other military air activities, including airlift operations. Military flights would, of course, receive maximum priority, perhaps using special corridors set up to handle the very large volumes of air traffic that any major conflict would undoubtedly generate. In peacetime, however, military transport flights are normally conducted along civilian air routes, in the interests of compliance with international air traffic control procedures. In principle, this offers many advantages but it also poses a number of problems for the airlift planner. For instance, there is a finite limit on the volume of aircraft that a particular air traffic control system can handle. With civilian air traffic in Western Europe now increasing by some 12 to 15% per year, that limit has already been reached in some sectors, leading to the introduction of flow control measures (ie, restrictions on the number of aircraft movements that can be permitted in any given period). Hence, even where overflight or diplomatic clearance is likely to be granted, the planner must not assume that the civil airway structure will have sufficient capacity to cope with the proposed flow of military aircraft during a major peacetime operation or exercise. Unless his airlift plan is fully coordinated in advance with the relevant air traffic control authorities, the planner may well have to accept constraints which, by restricting his proposed flow of aircraft, could significantly extend the time needed to complete an exercise or operation.

Air-to-Air Refuelling

In some circumstances, such as the mounting of an airlift operation over exceptionally long distances or where missions are flown to support the deployment of combat aircraft (using dual-capable tanker/transports) the planner may decide to use AAR. (See the example in Figure 3.5.) The availability and location of bases from which the tankers can operate may also have an important bearing on the selection of the route.

Staging Airfields

As a general principle, aircraft should not be refuelled, nor crews be changed, at a forward airfield during an intensive airlift operation. During such operations, turn-round times should be as brief as possible. This is especially important when the reception airfield is itself threatened or under actual attack, as occurred in Vietnam during the classic siege of the US garrison at Khe Sanh. For many weeks, air-landed and air-dropped missions flown by USAF transport crews provided the only lifeline between Khe Sanh and the outside world. In such critical situations, it is of course imperative that ground times be kept to the absolute minimum, but even when the reception airfield is not under immediate threat, the same principle holds good. The reception organisation must still attempt to off-load and process the incoming payloads as quickly as possible; additional responsibilities such as the organisation of refuelling and crew-changes are distractions which could jeopardise the main task. Should it be necessary to refuel *en route*, and if AAR is neither feasible nor practicable, transport aircraft will need to use a suitable staging airfield. Ideally, such airfields should be located as close as possible to the preferred track; otherwise, time and fuel penalties will be incurred as a result of the ensuing deviations. Even where a staging airfield is conveniently adjacent to the desired track, some time and fuel penalties

PLATE 5.7. The USAF has 60 KC-10 Extender tanker/transports which can dispense fuel to combat aircraft during long deployments while simultaneously carrying the latter's maintenance personnel and equipment. Seen here dispensing fuel to an F-15, the KC-10 is covered in more detail in Chapter 3.

are inevitably exacted by the need to descend, land, taxi, take off and climb back to cruising altitude. These considerations must be fully taken into account at the planning stage.

Diversion Airfields

In drawing up the overall airlift plan, the staff must give careful consideration to the availability of diversion airfields. In peacetime[4] a transport aircraft must carry sufficient fuel to enable it to divert to a suitable alternative should bad weather or some unexpected contingency, such as a blocked runway, preclude a landing at the destination. In some cases, however—when, for example, the destination is a remote island—the nearest alternative airfield may be so far away from the destination as to make it either impossible, or prohibitively punitive in terms of payload traded, to carry the large quantity of fuel needed to divert to such an airfield. In such circumstances, and provided that the destination is in a region of stable climatic conditions where poor weather is unlikely to persist for more than brief intervals, some air forces permit transport aircraft to carry 'holding' fuel in lieu of diversion fuel.[5] 'Holding' (or 'Island holding') fuel is usually sufficient to allow an aircraft to loiter in the vicinity of the destination for about an hour should this be necessary. After weighing such considerations as the destination's weather factor, number of runways and approach/landing aids, the planner must decide whether he can safely recommend the use of 'island holding' reserves for a particular operation. If so, it may be necessary to remind units of the procedures involved, notably the requirement to calculate and use a 'point of no return'.[6]

Crew Availability and Utilisation

In intensive airlift operations, the availability of aircrews may be a more limiting factor than the availability of aircraft. Depending on serviceability rates and maintenance intervals, military airlifters can keep flying for sustained periods whereas the work-cycle of their crews must be punctuated by regular periods of rest. It is therefore essential to ensure at the outset that the airlift plan is geared to the number of crews that are actually available. In assessing crew availability and utilisation, the planning staff must take due account of crew working periods. In peacetime, all air forces impose maximum periods of duty for transport aircrews in order to prevent excessive fatigue with its concomitant implications for flight safety.[7]

Although the maximum periods of duty for military crews (often 16 hours or more) are usually significantly longer than those observed by civilian crews, they may nevertheless prevent missions of longer-than-average duration from being undertaken by a single crew. In the case of a peacetime mission extending for more than, say, 16 hours the crew would normally be 'slipped' (ie, replaced) at a staging airfield. When it is impracticable to run a slip crew operation, which by definition requires the positioning and participation of additional crews, missions may need to be interrupted to allow crews to take a minimum rest period (typically 12 hours between landing and take-off) before continuing their flight. In cases where it is necessary to mount exceptionally lengthy AAR-supported missions, with no landings *en route* to the destination or in some instances between departure and return to base, the basic flight deck crew may be augmented by an extra pilot, navigator or flight engineer, hence allowing the primary occupants of these positions to be relieved for some part of the flight.

After completing particularly long flights, with or without augmentees, aircrews may sometimes have difficulty in sleeping during their subsequent rest periods, especially if they are suffering from time-zone disorientation. Special drugs, with no harmful after-effects, are now available for use in such circumstances and were successfully prescribed for some RAF C-130 crews during the 1982 Falklands campaign to ensure that they obtained adequate rest between long-duration missions.

Capacity of Reception Airfield

Having considered the dimensions and facilities of the main reception (or destination) airfield, the staff must address several further questions concerning its capacity before the overall airlift plan can be finalised.

Maximum Number of Aircraft on Ground

The maximum number and mix of aircraft types which can be on the ground simultaneously must be accurately assessed. Where the reception airfield has ample parking space, this figure is unlikely to prove restrictive unless the airlift is particularly intensive. Where space is limited—perhaps to the extent that only four or five aircraft can be accommodated on the ground at any one time—this will constrain the rate at which missions may arrive and hence extend the timescale required for deployment.

Other Limitations

The problem of airfield capacity is not merely a function of parking space. Airfields with small parking areas also tend to be limited in other ways; for example, they may have only one runway with no parallel taxiway and they may lack all-weather capability because they are not equipped with suitable approach and landing aids. These problems will certainly be compounded if the reception airfield is used as a permanent or temporary base for other aircraft which need to continue to operate throughout the airlift phase, thereby imposing further restrictions on air transport movements. The airlift planner must take full account of these factors and build in sufficient intervals between missions to allow some flexibility in coping with the inevitable early/late arrivals and departures.

Cargo Off-load Procedures

The time needed to off-load an airlifter depends on several factors including:

- Type of aircraft (eg, does it have integral ramps and built-in cargo-handling equipment?).
- Type and amount of cargo.
- Load configuration.
- Availability of handling equipment.
- Expertise of airloadmaster and aerial port personnel.

The loading and unloading of vehicles, pallets and other cargo from aircraft such as the B-707 and KC-10 (which do not have ramps) can be very time-consuming, even with good on-board facilities and a well-equipped, efficient aerial port team. Cargo handling is much easier when there is full-width access to the main hold via integral ramps (see Plate 5.9). With aircraft such as the C-5 and AN-124, which are equipped with spacious drive-through cargo decks, wheeled or tracked vehicles can be driven on board via one ramp and out of the hold via the other without any need for time-consuming reversing manoeuvres at either the departure or destination airfield. This facility is particularly beneficial in view of the large payloads carried by such aircraft. However, in the case of aircraft such as the C-130 and AN-12, which have only a rear ramp, vehicles and trailers are normally reverse-loaded at the point of departure so that they can be driven off rapidly on arrival at the reception airfield.

Engine-Running Off-loads

Time can also be saved by adopting engine-running offload (ERO) procedures. Used extensively by some air forces, including the USAF and RAF, the ERO (whereby aircraft engines are kept running at low power settings throughout the unloading sequence) can significantly increase the flow of missions through the reception airfield, and hence should always be considered at the planning stage. Provided that aircraft crews, marshalling personnel and aerial port teams are well briefed and preferably experienced in this procedure (particularly important if EROs are to be conducted safely at night) troops, vehicles and other equipment can be unloaded in a matter of minutes. For example, depending on the type of payload, time needed for taxying, and air traffic control clearances, a C-130 can land, complete an ERO and be airborne again within 10 minutes.

In hostile scenarios, rapid off-loads (and on-loads) are invaluable in reducing the vulnerability of both aircraft and payload. The Argentinians are thought to have used the ERO technique to good effect at Stanley Airport during the Falklands war of 1982. By operating under cover of darkness, and restricting their ground times to

PLATE 5.8. Unloading in progress through side freight door of KC-10 Extender. This aircraft does not have the rear doors and ramp found in most military airlifters.

PLATE 5.9.   C-5 with forward ramp in open position.

just a few minutes, the Argentine Air Force transport aircraft used to supply the beleaguered garrison were completely successful in avoiding detection and attack by nearby units of the British Task Force. Rapid off-loads were also the order of the day at the height of the Dhofar campaign in Oman during the early 1970s when the main Omani airhead at Salalah was frequently under attack from rebel artillery and mortar fire.

Even in less critical situations, EROs are valuable in reducing ground time, easing congestion on the parking ramp, and injecting an extra degree of urgency into the proceedings. The ERO technique—which can equally well be used for rapid on-loads—also reduces the risk (albeit slight) of aircraft being unable to restart one or more engines after shutting down at an airfield which may have little or no maintenance or rectification facilities. On the other hand, EROs are not advisable with payloads that are awkward and time-consuming to unload, such as heavy unpalletised freight. As a rule of thumb, if an off-load or on-load is expected to take more than about 15 mins, it would normally be preferable to shut down the engines.

Runway Off-loads

At some smaller airfields, it may be necessary for larger aircraft (such as the AN-124 and C-5), which may be unable to manoeuvre onto the taxiways or parking ramp,

PLATE 5.10.   Vehicle and trailer reversing from ramp of Luftwaffe C-160.

to remain on the runway for off-loading. Where there are two or more runways, this need not present too many problems, but where there is only one runway this measure would effectively close the airfield to other traffic during the period required for unloading. Whereas vehicles can be rapidly driven out of the holds of these large aircraft, the unloading of pallets and other freight can be a time-consuming process, sometimes requiring 2 to 3 hours. Hence, where there is no alternative to handling large aircraft on the active runway, the planner must either schedule such missions for periods when the flow-rate is at its lowest, or substitute smaller airlifters if available and able to carry the type of payload in question.

Taxying Off-loads

In extreme circumstances—where, as in the examples quoted earlier, the destination airfield is under actual or imminent attack, but where reinforcements or supplies must be air-landed despite the high risks—aircraft might have to discharge their payloads while still taxying. Though obviously unsuitable for some types of cargo, the technique involved is crude but effective, requiring the use of aircraft with a rear ramp which would be opened, once the aircraft had braked to a safe speed, to allow the airloadmasters to push containers of food, ammunition and other supplies over the sill for retrieval by the receiving unit. This was but one of several unorthodox measures used to resupply the US garrison besieged at Khe Sanh in 1968.

PLATE 5.11.   Arctic vehicle and trailer emerging from RAF C-130.

Flow Plan

After carefully considering and weighing all of the factors described in this chapter, the airlift planner's final task is to produce and publish a flow-plan—designed to ensure an even flow of missions along the route—for each phase of the exercise or operation (ie, deployment, resupply and redeployment). This series of flowplans

PLATE 5.12.   Example of a load which does not lend itself to the ERO technique.

PLATE 5.13.   ERO from C-130.

should provide detailed information on each flight in terms of aircraft type, mission number, itinerary (ie, departure, staging and destination airfields), estimated times of departure and arrival, total elapsed time of each mission, and payload. Depending on the size and complexity of the airlift operation, the flowplan might be rather a daunting document, requiring considerable care and time in its preparation. However, once the basic facts and parameters are assembled, the final flowplan can be readily produced by computer always provided, of course, that the planner has access to the necessary hardware, software and human expertise. When published, the flowplan serves as the blueprint for the airlift operation, defining its pattern in terms of aircraft employed, mission rate, and timescale. Armed with this information, the various commanders, flying units and supporting organisations should have a clear picture of the plan which they will collectively be required to implement.

Questions

1. What factors govern an airfield's ability to accept various types of transport aircraft?

2. (a) What is the purpose of the LCN/LCG system?

   (b) Under what circumstances can an airfield in LCG category III accept operations by an aircraft with an LCN of 80?

3. Name three types of support facilities needed by an airfield if it is to sustain a major airlift operation.

4. (a) What is the basic aim of performance planning?

   (b) Define the four stages into which flights are divided for performance planning purposes.

5. Explain what is meant by a 'WAT Limit'.

6. List five key mission planning factors which determine the overall airlift plan.

7. What considerations govern the time needed to off-load a transport aircraft?

8. What are the chief advantages and disadvantages of engine-running off-loads?

6

Air-Drop Operations and Techniques

INTRODUCTION

The employment of air and ground forces on air-drop operations can be extremely controversial. This was certainly so during the Second World War in which assaults by air-dropped units achieved widely varying results. For example, while Germany's bold use of paratroops to seize Crete in 1941 (Operation MERKUR) was largely successful, the Allied airborne operation in The Netherlands in 1944 (Operation MARKET GARDEN) conspicuously failed to accomplish its objective, despite the heroism of the aircrews and ground forces involved.

The controversy surrounding the air-drop concept has, if anything, intensified in recent years. Certainly the survivability of transport aircraft when threatened by modern air defence systems, coupled with the vulnerability of lightly equipped paratroops once on the ground, make air-drop operations a high-risk undertaking today, particularly in the Central Region of Europe. That said, many analysts would concede that the rapid deployment of even a small parachute force can—if adroitly exerted at a critical time and place—achieve results out of all proportion to the potential risks. Their capacity to strike a sudden, decisive blow (perhaps by surprising the enemy in a location where he least expects an attack) gives air-drop operations a unique potential which is reflected in the special *esprit de corps* of both the air and ground units involved. Hence it is hardly surprising that, whatever the advantages and disadvantages, the airlift of troops and equipment into forward airfields or strips (defined as the air-land option because the transport aircraft actually land with their payloads) is seldom perceived to be as challenging or exciting as their delivery by parachute onto a dropping zone (DZ).

Yet whilst it is true that the ability to conduct air-drop operations is important to the wider employability and credibility of a tactical transport force, the fact remains that air-drop missions usually form a relatively small proportion of the overall airlift task. This is in line with the general principle that air-drop operations should not normally be conducted if it is feasible instead to mount a conventional airlift into a suitable airhead close to the objective. There are several reasons for this:

- The air-land option offers both air and ground commanders greater flexibility in employing the total forces at their disposal. Even in large armies, only a relatively small percentage of combat and support troops are parachute-equipped

123

PLATE 6.1. C-141B Starlifter dropping equipment platforms.

and trained, while not all tactical transport crews may be trained or available for air-drop operations[1].

- The air-land option avoids the injuries to personnel and damage to vehicles which must be expected during a parachute drop as a result of heavy landings and equipment malfunctions. The exact degree of attrition will vary according to such factors as surface wind velocity, light conditions, type of DZ surface and dropping height, but as a general guide the ground commander would normally expect about 3% of paratroops to be injured during a daytime drop with surface wind speeds of up to 13 kt. By night, this percentage would probably increase to about 5%. As wind speeds increase beyond 13 kt, so the percentage of injuries would also rise until, above the upper limit of about 18 kt, the projected casualty rate would probably be deemed unacceptable. In short, the weather factor is much more critical in air-drop missions than in air-landing operations.

- Paratroops may become scattered and separated from their heavier equipment, especially if different, albeit adjacent, DZs are used for personnel and material. This makes paratroops particularly vulnerable immediately after landing when, as well as regrouping, they must locate and de-rig equipment platforms and bring their support weapons into action as soon as possible. Even a well trained battalion may need 20–30 min by day and as much as 40–50 min by night to reorganise. Air-landed units do not have this problem of dispersion.
- Provided that a suitable airhead is available, air-landing is a much more efficient way of deploying and resupplying a force than delivery by parachute. Since no space or weight is needed for despatch crews, parachutes and equipment platforms, more troops and cargo can be carried for a given number of missions. For example, an unstretched Hercules can air-land 92 fully equipped combat troops but can air-drop only 64 paratroops. Similarly, the stretched C-130 can carry a mixed load of three long-wheel base Land-Rovers, three 105-mm pack-howitzer guns and 20 troops but, in the equipment mode, can air-drop only two medium-size platforms owing to centre-of-gravity considerations.

AIR-DROP TASKS

Despite the desirability of using the air-land option whenever possible, there are particular missions in both war and peacetime scenarios where the application of air-drop techniques can be the best if not the only way to achieve the desired result. Some of the more important of these tasks are listed below, classified according to their likely context.

Large Scale War

Use of paratroops or special forces (eg, Soviet *Spetsnaz* units) to:

- Seize vital ground and/or key facilities, including airfields.
- Capture strategic targets in enemy-controlled territory.
- Destroy enemy command, control and communications systems.
- Attack enemy reserves.
- Disrupt logistic and other activities in enemy rear area.
- Counter-attack enemy penetration of own rear area.
- Reinforce friendly forces, especially on flanks.

Limited War

Use of paratroops or special forces to:

- Influence events without actually intervening by adopting overt 'stand-by' posture.
- Seize a point of entry (such as an airfield, port or beachhead) for subsequent exploitation by air-landed or amphibious forces.
- Mount a *coup de main* operation (eg, seize a strategically positioned bridge).

- Provide rapid reinforcement of friendly forces.
- Reconnoitre and report enemy activity.
- Attack command, control and communications facilities, POL supplies, weapon storage compounds, railways, etc.
- Rescue or protect expatriate nationals caught up in hostile situations overseas.

Peacetime

Dropping of supplies to support:

- Isolated garrisons where weather or lack of suitable airfield precludes conventional airlift.
- Humanitarian relief operations in inaccessible areas.

ADVANTAGES AND DISADVANTAGES OF AIR-DROP OPERATIONS

Before examining how tasks such as those outlined above are planned and executed, it is important to examine in more detail the advantages and disadvantages of air-drop operations and to consider how the latter can be minimised. In principle, a well-trained parachute force—that is to say, a combination of specialist tactical air transport units and paratroops—is ideally suited to the rapid and flexible projection

PLATE 6.2. Luftwaffe C-160 free-dropping food supplies in Ethiopia in 1985.

of military power. It can be rapidly assembled at a secure base, and delivered to its destination—independently of fixed entry points such as ports or airfields—over strategic distances (using staging-posts or AAR) and across virtually any terrain. Furthermore, a parachute force is well placed to exploit the element of surprise; after a low-level transit to reduce the risk of detection, a formation of 15 Hercules aircraft can deliver a battalion group of over 600 paratroops, complete with vehicles, support weapons and ammunition, onto two adjacent DZs in little more than 5 min. The mere possession of such a capability poses a threat which no potential adversary can afford to ignore, and which helps to explain why parachute forces (sometimes referred to as 'airborne forces') feature so prominently in US and Soviet military doctrine. If committed sufficiently early, paratroops can also be used to stifle or at least contain a 'brush fire' conflict before it can develop into a major conflagration. The Franco/Belgian action at Kolwezi in Zaire in 1978—albeit mounted primarily to rescue a large group of expatriate hostages—was a good example of a successful intervention operation. Rebel forces holding the area were taken completely by surprise when a battalion of French paratroops was dropped onto the local airfield, which was then secured for the subsequent air-landing of Belgian troops. Confronted by this sudden projection of force, the rebels quickly capitulated and most of the hostages were evacuated unharmed.

As already indicated, however, committal of a parachute force frequently entails a high degree of risk. Quite apart from their limitations *vis-à-vis* conventional airlift missions, air-drop operations—like most transport operations—are normally feasible only if the enemy does not enjoy a favourable air situation in the vicinity of the route and DZ[2]. Otherwise, such is the vulnerability of a stream of low-flying transport aircraft to air- or ground-based defence systems (particularly during the final run-in to the DZ when the aircraft must climb up from low level to their dropping height) that even with fighter, ECM and SEAD[3] cover, the likely rate of attrition would probably be unacceptably high. Depending on the sophistication of the enemy's air defences, this risk can be partially alleviated by:

- Mounting the operation at night (though this may make it more difficult to acquire the DZ unless a pathfinder team has been inserted earlier to provide visual or electronic markers).
- Flying as low as possible (say, not above 200 ft) *en route* to and from the DZ.
- Dropping personnel and equipment at the lowest possible height subject to practical limitations (eg, operational parameters of parachutes).
- Using tactical routes, selected to avoid known enemy positions and to exploit maximum cover from terrain.
- Maintaining radio and radar silence.
- Operating in small elements along separate routes, merging into a single stream of aircraft only shortly before the final run-in to the DZ.

Even if there is no significant air threat, air-drop operations remain subject to several further limitations. As noted earlier, parachute assaults cannot take place when the surface wind at the DZ is more than about 18 kt, at least not without risking an unacceptably large number of casualties. *En route* weather is also important; unless equipped with INS and station-keeping equipment (discussed later), a stream of aircraft encountering poor visibility and/or low cloud may be forced to break formation or make large deviations from track. The limited number of weapons, vehicles and supplies that can be dropped during a given operation may also pose serious problems, particularly for smaller air forces which do not enjoy the air-dropping capability of the USA and USSR. With priority having to be given to weapons (such as mortars, artillery, SAM and anti-tank missiles) it may be impossible to drop as many vehicles as the paratroops would like, thereby denying them full tactical mobility. That being the case, DZs should be as close to the initial objective as possible. Finally, it is important to remember that, unless it proves possible to seize and secure an airfield for subsequent air-land operations, it may prove extremely difficult to resupply—much less extract or redeploy—a parachute force once it has been committed.

PLANNING FACTORS

Having decided that a parachute operation is essential, feasible and incapable of being replaced by an orthodox air-land operation, the air and ground force staffs must draw up a series of detailed plans, beginning with the post-drop tactical plan— to which all other plans must ultimately be geared—and then working backwards through the assault and airborne phases to the initial mounting plan. Hence the planning sequence is as follows:

PLATE 6.3. Team preparing to despatch supplies by free-drop method from C-130.

Ground Tactical Plan

This is the basic army plan and must include:

- Objectives.
- Order of battle and weapons mix required.
- Requirement (if any) for fire support (eg, from combat aircraft, warships or artillery).
- Provision of mobility.
- Requirements for resupply and reinforcement.

Assault Plan

This is the joint army/air plan which covers all aspects of the air-drop operation, including:

Selection of DZ

As this is directly related to the overall tactical plan, the ground force commander is responsible for identifying potential DZs. However, given the air force's responsibility for delivering the force, the final selection of DZs is usually a joint decision based on such factors as:

- Proximity to objectives.
- Enemy dispositions (including air defences).
- Availability of large, clear areas with suitable dimensions and characteristics for air-drop of both personnel and equipment.
- Proximity to terrain features such as mountains, woods or rivers.
- Proximity to airfields which could be seized and secured for follow-up operations.
- Ease of recognition from the air.

Selection of 'P' Hour

'P' Hour (the time at which the drop or assault is set to commence) may be in daylight, twilight or darkness depending on the overall air and ground tactical situation. Many air and ground commanders favour an assault at first light. This offers the best chance of avoiding detection *en route* to the DZ and catching the enemy off guard, whilst allowing the paratroop units to regroup and move out to their initial objectives in daylight.

Use of Pathfinders

The chances of a successful air-drop operation can be much enhanced by inserting pathfinders—typically a four-man team of special forces soldiers—ahead of the main assault to locate, reconnoitre and mark the respective personnel and equipment DZs. In addition to providing visual or electronic markers to guide the aircraft during their final run-in, pathfinders may be able to assist the aircrews in achieving an accurate and concentrated drop by transmitting details of the surface wind velocity and atmospheric pressure as the force approaches the target area. For obvious reasons, it is imperative that the positioning or presence of pathfinders should not compromise the main operation. To minimise this risk, the team would probably be inserted 24 to 48 hours ahead of the main assault using various clandestine techniques, such as free-fall parachuting by night into an area some 20 to 30 miles from the assault DZ, or infiltration over land after initial deployment by helicopter or submarine.

Selection of Target Approach Point (TAP)

Ideally, a TAP should be a prominent geographical feature (eg, a headland or river bend) about 10 nm before the DZ. Its function is to provide one last check-point at which the timing and integrity of the formation can be confirmed before commencing the final run-in. The TAP should be selected not only for its ease of visual recognition, but also to ensure that the track from the TAP runs down the longest axis of the DZ, if possible with the azimuth of the sun behind the aircraft. Selection of the TAP must also, of course, take due account of enemy dispositions.

Air Plan

As the name indicates, this is the air OPLAN which includes the routes, heights and speeds to be flown, and the procedures and tactics to be adopted. During the period from take-off to arrival over the DZ, overall command of the operation is usually exercised by the senior air commander, but the final decision as to whether the operation is to proceed, be postponed or be aborted must be taken jointly by the air and ground commanders.

Mounting Plan

This is an army-produced plan (with some air inputs) which covers arrangements for the assembly and preparation of the parachute force and its equipment at the mounting base and airfield. It includes details of individual aircraft loads and establishes deadlines by which the various stages in the pre-launch sequence must be completed.

AIR-DROP TECHNIQUES

Paratroops

Paratroops usually jump in simultaneous 'sticks' from twin para-doors at the rear of the aircraft, their canopies being opened automatically by static lines attached to the aircraft. Dropping troops in parallel ensures optimum concentration on the ground and minimises the length of DZ required. Even so, a stretched C-130 dropping its maximum of 92 troops (in simultaneous sticks of 46) will require a DZ which is approximately 10,000 ft × 1,200 ft (3,050 m × 366 m). Speeds and heights vary with aircraft type and national procedures, but a C-130 would typically fly at 120 kt and 800 ft above the DZ during the dropping sequence. To achieve this highly vulnerable profile, the aircraft would have to 'pop up' from low level and slow down to dropping speed about 4 nm before the DZ. Experiments have been conducted to establish whether paratroops could be safely dropped from much lower heights using multi-canopy parachutes, with a view to reducing or even eliminating the need for 'pop up'. However, the results have not been encouraging. For example, a heavily laden parachutist (who, with full equipment could weigh anything up to 310 lb) would have only some 8 to 10 sec between leaving the aircraft at 250 ft and arriving on the ground. This would leave insufficient time to check his canopy, carry out vital drills, lower his equipment container and prepare for landing, much less operate his reserve should his main parachute have failed.

In the final analysis, the vulnerability of an aircraft and all its occupants must be weighed against the potential risks to the paratroops entailed in an exceptionally low-level drop. In the interests of striking a sensible balance between these factors, a paratroops commander would probably accept a drop height of 500 or 600 ft if operationally justified. At the same time, he would have to point out that the lower the drop height, the greater the proportion of casualties caused by heavy or uncontrolled landings.

PLATE 6.4. US paratroops emplaning on C-130. Note the size of their equipment and weapon containers.

Special Forces

Precise details of the parachuting methods used by special forces (SF) are of course highly classified in all countries. However, it is known from a variety of open sources that their basic *modus operandi* is to employ the high-altitude/low-opening (HALO) technique. Using aircraft such as the C-130, which would of course need to be depressurised prior to the drop, a team of perhaps 12 SF soldiers, equipped with oxygen, would prepare to jump from the rear ramp. After leaving the aircraft at, say, 20,000 ft they would free-fall initially before deploying custom-built 'flying' canopies which would be steered to clandestine landing spots up to 20 nm from where the initial drop occurred. Once inserted, SF or other forces can be resupplied covertly using either conventional or HALO air-drop techniques. In the latter case, the parachutes would probably be opened at about 1,000 ft by pre-set barometric devices. Some air forces maintain a small number of specially trained tactical transport aircrews (and in some cases specially equipped aircraft) to support SF air-drop operations.

Equipment (Normal Techniques)

Using normal techniques, equipment platforms, containers or packs are usually dropped at lower heights than personnel. For example, an RAF C-130 would normally drop a medium-sized platform (carrying, say, a Land-Rover and trailer) from 600 ft,

PLATE 6.5.   Paratroops of the Federal Republic of Germany's First Airborne Division
jumping from Luftwaffe C-160.

a batch of 1-ton containers from 500 ft, and small single packs from 250 ft, all at speeds of about 120 kt. As a general rule, the lower the drop height, the greater the degree of accuracy that can be expected. Like the height, the dropping procedure itself varies with the type of load. Heavy platforms are usually extracted from the cargo hold by a series of small parachutes (activated from the flight deck) which pull the load clear of the aircraft before releasing the main canopies for the descent. One-ton containers, on the other hand, are usually ejected manually with the aircraft being flown in a slightly nose-up attitude to ease the task of the despatch crew in manoeuvring the loads over the roller-conveyor system and rear sill. Although there are differences in national procedures, the techniques described here are fairly standard on aircraft such as the AN-12, C-130, C-160 and G-222.

Equipment (Special Techniques)

When equipment must be delivered with a very high degree of accuracy, or where operational circumstances preclude the use of normal methods, a number of special techniques can be employed.

Low Altitude Parachute Extraction System (LAPES)

LAPES is a demanding but highly effective method of delivering heavy loads with great precision either onto airfields where it would be impossible or imprudent to land, or onto any suitable flat area.

As illustrated in Figure 6.1, the aircraft is flown (with landing gear down in case of momentary contact with the ground) at a height of between 5 and 10 ft above the ground, at a speed (in the case of the C-130) of 120 kt. At a pre-designated point, marked on the ground by the reception team, a crew member releases the extraction parachute. Some 3 sec (or 600 ft of ground travel) later, the load leaves the hold, coming to rest about 750 ft after activation of the parachute. Ground slide averages between 100 and 200 ft. One 28-ft parachute will extract loads weighing up to 12,000 lb; two parachutes are used for loads between 12,000 and 24,000 lb, three for loads between 24,000 and 36,000 lb and so on. Loads of up to 50,000 lb have been successfully delivered by USAF C-130s using the LAPES technique, which is also used by a number of other air forces.

Parachute Low Altitude Delivery System (PLADS)

Using PLADS, an aircraft can deliver individual 1-ton containers to within about 100 ft of the desired impact point. The method used is similar to LAPES except that the drop is executed from 200 ft above the DZ.

As illustrated in Figure 6.2, the aircraft (in this example, a C-130 flying at 120 kt) approaches the DZ at 200 ft with the extraction parachute reefed and trailing in the slipstream. Using a special sighting device, the co-pilot activates a de-reefing mechanism when the aircraft reaches a pre-designated point, defined by reference to markers positioned near the DZ by the reception team. This causes the parachute to inflate, thus extracting the load which follows a predicted trajectory to the target.

PLATE 6.6. Conventional drop of heavy equipment by C-141B.

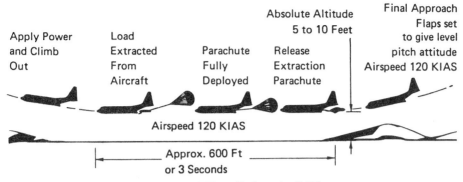

Apply Power and Climb Out	Load Extracted From Aircraft	Parachute Fully Deployed	Absolute Altitude 5 to 10 Feet Release Extraction Parachute	Final Approach Flaps set to give level pitch attitude Airspeed 120 KIAS

Airspeed 120 KIAS

Approx. 600 Ft or 3 Seconds

FIG 6.1. LAPES profile flown by C-130.

PLATE 6.7. LAPES-delivered jeep using single parachute. Note that the landing gear is down.

PLATE 6.8. LAPES-delivered armoured vehicle, using three parachutes.

PLATE 6.9. Side view of armoured vehicle leaving MAC C-130 which is already beginning to climb away after completing the LAPES manoeuvre.

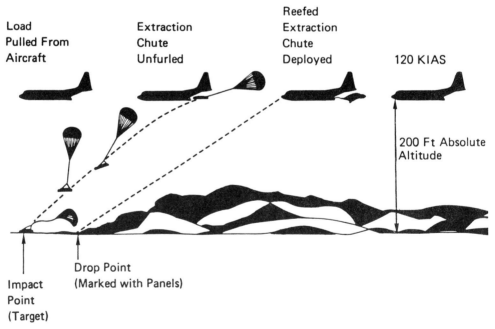

Load
Pulled From
Aircraft

Extraction
Chute
Unfurled

Reefed
Extraction
Chute
Deployed

120 KIAS

200 Ft Absolute
Altitude

Drop Point
(Marked with Panels)

Impact
Point
(Target)

FIG 6.2. PLADS profile flown by C-130.

Horizontal velocity is negligible just before touch-down. Hence there is very little slippage on the ground.

Container Delivery System (CDS)

This system has been specifically developed by the USAF to facilitate the rapid and concentrated dropping of large numbers of A-22 resupply containers from the C-130. CDS is more accurate than normal air-drop methods, but lacks the precision of LAPES and PLADS. The A-22, measuring $4 \times 4.5 \times 4$ ft and capable of holding up to 2,200 lb of cargo, is the US equivalent of the British 1-ton container. Mounted on simple, plywood platforms, which are rigged on the cargo hold roller conveyor system, 16 of these containers can be carried—and dropped in a single pass—by the C-130. The options are very flexible. Containers can be dropped one at a time or up to eight in a single stick; alternatively, they can be rigged two abreast in the hold and dropped in pairs up to a maximum of two simultaneous sticks of eight. The Hercules approaches the DZ at 500 ft and 130 kt, with a $6°$ to $8°$ nose-up attitude. At the pre-computed air release point, the navigator operates a mechanism which shears the nylon aft restraint gate, allowing the load to roll backwards down the roller conveyors and over the rear ramp by the force of gravity. On leaving the aircraft, the parachutes are automatically deployed by static line.

Low-Level Free Drop

The 'free drop' method is exactly what the term suggests: supplies are pushed out of the aircraft without parachutes as it flies low and slow over the DZ. Although the system has obvious limitations, being feasible only when the supplies in question are sufficiently robust or resilient to withstand the considerable impact involved, it also has several advantages. For example, it is ideal for dropping bulk food such as grain, rice and flour to remote garrisons or communities where the absence of airstrips and surface access makes air delivery essential. Free-dropped supplies are usually packed in strengthened double-thickness sacks (in order to minimise the risk of spillage or bursts) which are then mounted on disposable plywood bases in sizes and weights which can be readily manhandled out of the aircraft as it crosses the DZ. This method is very cheap, unlike techniques involving the use of parachutes which, over a sustained operation, can become extremely expensive especially if, as is often the case in such locations, they cannot be retrieved for use in subsequent drops. For the aircrew and despatch team, the free drop procedure is extremely simple. With landing gear selected up or down (depending on national operating procedures) the aircraft approaches the target at about 120 kt and a height of 50 ft and, at a pre-arranged point (which may or may not have been marked on the ground) the co-pilot or navigator gives a warning followed 2 to 3 sec later by an executive command to the despatchers to eject the load. Several runs across the DZ would be necessary to drop all the supplies which could be delivered in this manner by aircraft such as the C-130 and C-160. Ground slide is usually about 150 ft. The free drop technique was used with conspicuous success over a period of several months by aircraft taking part in the international relief operation in Ethiopia in 1985.

STATION KEEPING EQUIPMENT

In order to have the best chance of success, a parachute assault operation needs to be as concentrated as possible in both time and space. For example, no more than 5 min should be taken—if the operation is to achieve maximum surprise—to complete the drop of a battalion of about 550 troops plus weapons, ammunition and vehicles. To achieve this timescale, the stream of aircraft involved (typically 15 C-130s for a battalion-scale assault) would need to cross the DZ at fairly close intervals, which in turn means that they would also have to be flown in a cohesive and closely coordinated formation during the transit from the mounting base. This does not mean that the stream of aircraft must operate as a single formation throughout the mission. On the contrary, there would be tactical advantages in sub-dividing the force into smaller elements *en route* to and from the DZ. Thus, an overall force of 15 aircraft might be split into three elements of five, each flying its own route to the TAP. Nevertheless, the integrity of each element would have to be maintained continuously from take-off and the various elements would need to merge into a fully integrated stream at the TAP.

Flying at horizontal intervals of 4,000 ft—a spacing commonly adopted by tactical transports engaged on such operations—a stream of 15 aircraft will extend some 10 nm. With accuracy of timing and navigation of crucial importance, adjustments in heading and speed will be inevitable, yet difficult to synchronise across the whole stream, especially if radio silence is being observed in order to reduce the risk of detection. Hence, even when weather conditions are favourable, the difficulty of commanding and controlling a large low-level formation of tactical transports should not be under-estimated. When the weather *en route* deteriorates to the point where aircraft lose visual contact with one another, the problems faced by stream, element and individual aircraft commanders are increased by several orders of magnitude. If

PLATE 6.10.   Luftwaffe C-160 delivering food supplies by free drop in Ethiopia. Sacks in the foreground, for the most part intact, were dropped on the previous run.

such conditions persist for more than 1 to 2 min, the formation will almost certainly have to disperse according to a prearranged plan designed to ensure safe vertical and horizontal separation. Once scattered in this way, it is very difficult for the aircraft to regain formation, failing which the operation would have to be aborted.

As with other applications of air power, the need for an all-weather parachute assault capability has assumed a much greater importance in recent years. With this in mind, a US avionics company (Sierra Research) has produced and marketed a secondary radar system appropriately termed Station Keeping Equipment (SKE). Use of this system, which is now in service with the USAF and RAF, enables up to 36 aircraft to establish and maintain a safe formation even when they lose all visual references and contact. SKE thereby not only renders parachute operations significantly less susceptible to bad weather, but also reduces the risks of such operations when conducted in marginal weather or at night. The SKE system currently used by the RAF Hercules force is the AN/APN-169B, which provides 360° cover over a range of 10 nm.

In the RAF configuration, most components are located on or near the pilots' instruments panels (see Figure 6.3) although two (a transmit/receiver and coder/decoder) are in the cargo hold. The equipment has two aerials, an omni and a directional antenna, both fitted to the underside of the fuselage. The plan position indicator

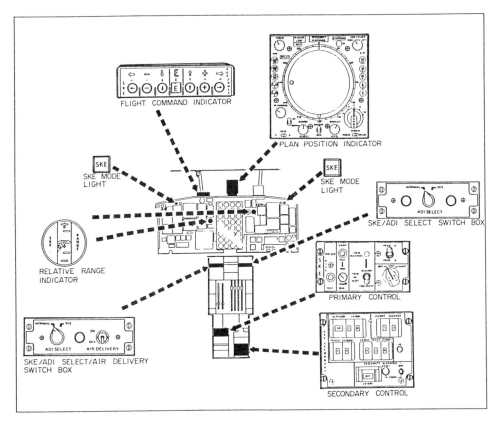

FIG 6.3. SKE instruments and controls on flight deck of RAF Hercules.

(PPI) (or radar scope) is located mid-way between the pilots on the flight-deck coaming; this is the primary SKE display, showing the range and azimuth of other aircraft in the formation. The equipment is operated from the primary control panel situated to the left of the centre pedestal. The secondary control panel (at the rear right-hand side of the centre pedestal) operates the 'track while scan' (TWS) and proximity warning systems. The TWS enables the pilot to fly to a set point in space behind another aircraft, with the SKE measuring the slant range and relative angle before converting this data into along and across track components. These are compared with preset values and any difference is fed to the steering bar on the flight director and the relative range indicator. Should any aircraft in the formation encroach upon the pre-selected distance set on the secondary control panel, a horn will be triggered by the proximity warning system, alerting the crew to take corrective action. A flight command indicator (FCI) in front of the pilot allows a formation leader to send manoeuvring instructions to his formation without needing to broadcast verbal orders by radio.

A typical SKE formation consists of 15 aircraft, sub-divided into five elements of three. As illustrated in Figure 6.4, the second and third aircraft in each element are respectively displaced 500 ft to starboard and port of the element leader, with a longitudinal gap of 4,000 ft between each aircraft throughout the formation. Each element leader controls his section of three aircraft by means of his FCI, while the stream commander controls the overall formation by transmitting FCI instructions to element leaders in succession. When climbing, descending or adjusting speed, the formation is required to react in unison to an executive command, but changes in heading of more than a few degrees require an appropriate time delay to ensure that all aircraft turn at approximately the same point in space. About 3 min are needed to get a formation of 15 C-130s through a turn at their *en route* speed of 210 kt.

Blind Drop Facility

After using SKE to achieve a successful transit, a formation leader should not need to abort his mission in the final stages because of poor weather at the DZ. The SKE system therefore incorporates a blind drop capability. This involves the use of a portable ground beacon (or zone marker) which transmits signals during the approach to the target, providing positive identification of the DZ as well as bearing and range. Used in conjunction with the standard SKE information, this terminal guidance facility allows a formation of tactical transports to deliver their payloads 'blind'—ie, without visual reference to either the ground or each other. However, it may be difficult to establish the ground beacon on the DZ in advance of the assault. This is a classic task for pathfinders or special forces, using the techniques described earlier, but the utmost care must be taken not to compromise the operation through either physical or electronic detection by the enemy.

Limitations of SKE

In addition to the problem of getting a ground beacon into position on a DZ deep inside enemy territory, SKE is subject to several other limitations. The most serious operational drawback is the equipment's distinctive radar 'footprint' which might

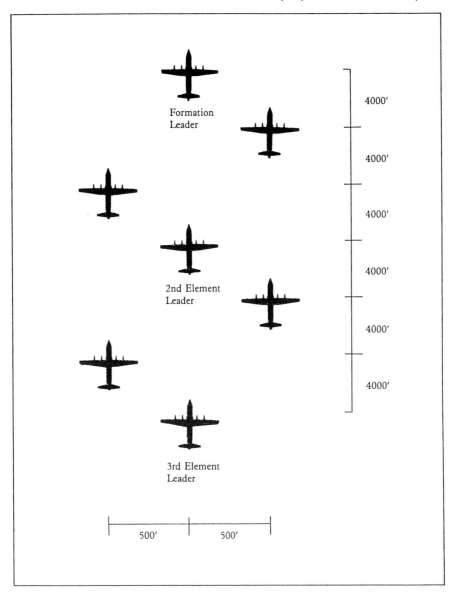

FIG 6.4.   Typical SKE formation.

preclude its use in a hostile environment. As well as being highly vulnerable to electronic detection—and, incidentally, emitting a radar 'signature' which could be used to provide guidance to enemy air- or surface-launched missiles—the system is also highly susceptible to jamming. Finally, it is important to remember that, while SKE can be extremely effective in certain scenarios, it is only what its name suggests: a system which allows a group of aircraft to remain in formation irrespective of visual references. Apart from giving range and bearing from the ground beacon on the DZ, SKE provides no positional information and does not of itself permit aircraft to navigate in formation along a pre-determined track. In order to operate safely at very low level, below safety height, the crew of a tactical airlifter needs to know their exact position at all times. When weather conditions permit, the aircraft's position is determined and continuously monitored by a combination of map reading and interpretation of on-board navigation aids such as INS. In adverse weather, the crew must rely solely on their navigation equipment. It follows that this equipment must be extremely accurate and reliable if all-weather low-level operations are to be undertaken with any reasonable degree of confidence. In short, transport aircraft need high quality INS as well as SKE if they are to achieve the requisite degree of precision for all-weather missions in formation at low level. Without an effective INS, SKE's only contribution to a particular operation may be to enable the entire formation to stray off track together.

ILLUSTRATION OF A PARACHUTE ASSAULT OPERATION

Having outlined the principles and techniques of air-drop operations in some detail, this Chapter concludes with an examination of how some of these can be applied in practice by describing an imaginary but realistic parachute assault operation—Operation TALON. The setting is entirely fictitious but the execution of the mission is based on current operational concepts and procedures:

Operation Talon

Background Situation

A rebellion has occurred in a Third World state which is a major supplier of important raw materials to the West. In an attempt to form a breakaway state, a faction of the army has mutinied and seized control in the province where most of the known mineral reserves are located. The democratically elected government has therefore called upon one of its Western allies (with whom it has defence links dating back to the colonial era) to assist in ejecting the rebels from the provincial capital, the main access to which is by air. The rebels hold the local airport—the only airfield of any strategic significance in the province—but have no air defence aircraft or weapons. However, they have blocked the runways with vehicles to prevent an air-landed operation by government forces. Having no air-drop capability of its own, the government has readily agreed to a suggestion from its ally that the latter should mount a parachute assault to dislodge the rebels from the airport in order to provide an entry point for follow-up operations by loyal troops. Within 48 hours of the decision to proceed, the Western ally has assembled the necessary forces and aircraft (15 C-130s for the operation itself and two reserves) at the host nation's main military airbase near the Federal capital. Intelligence assessments indicate that a battalion of 576 paratroops should be sufficient to confer a 4:1 numerical advantage over the rebel infantry company of about 130 men thought to be deployed at the provincial airport.

Operation Order

In real life, a comprehensive Operation Order would be produced for such an operation, covering all aspects of the mission in great detail. In this fictional scenario, however, the much-abbreviated outline which follows should suffice to indicate the type of information which needs to be covered.

Enemy Forces. Approximately 130 infantry deployed in vicinity of DZ, equipped with personal weapons, light machine guns and mortars.

Friendly Forces.

- **Land:** One parachute battalion (576 paratroops equipped with personal weapons, heavy machine guns and mortars).
- **Air:** 15 SKE-equipped C-130s.

Mission. To assault, seize and hold the rebel-held airport until relieved by air-landed government forces.

Main Elements of Assault and Air Plans

- **DZ**
  - Airport 10 nm south-west of provincial capital.
  - Nine C-130s to drop personnel (64 from each aircraft) onto grassed area 200 metres east of main runway.
  - Six C-130s to drop equipment and vehicle platforms on taxiway 200 metres west of main runway.
  - Pathfinders/DZ marker beacon not required.

- **TAP**
  Prominent river bend, 12 nm south of DZ.

- **P Hour**
  First light (0400 Z).

- **Weather Forecast**
  - En Route
    Thick cloud in mountain areas, with ceiling down to 1,000 ft in places. Otherwise clear skies.
  - **DZ**
    Scattered cloud, tops 10,000 ft, base 1,000 ft. Visibility 5 nm. Surface wind velocity light and variable.

- **Duration of Air Mission**
  Mounting airfield – DZ: 2 hr 25 min
  DZ – Mounting airfield: 2 hr 15 min
  Total flight time: 4 hr 40 min
  - Route
    See Figure 6.5.

Narrative Description of Air Mission

After a stream take-off at intervals of 15 sec from the mounting airfield, the 15 C-130s take up their stations in the formation as quickly as possible, using SKE throughout the climb to flight level 200. As there is no operational reason in this scenario to operate at low level until crossing the provincial boundary (see Figure

6.5) the stream commander has elected to fly the first part of the mission at normal cruising altitudes. This saves fuel, provides a more comfortable ride for the paratroops and despatchers, and is less taxing on the aircrews. Nevertheless, full concentration is still required to maintain intervals of 4,000 ft between aircraft using a combination of SKE and, whenever possible, visual cues. The speed flown by the formation leader during the high-level phase is the normal speed appropriate to that particular combination of weight, altitude and temperature, reduced by about 15 kt to allow aircraft at the rear some flexibility in speed adjustments. During this initial segment (which is flown along an established airway) the SKE and navigation systems are carefully checked and the timing is closely monitored. Overhead the 'LAC' VORTAC, the stream commander orders the formation to descend at 230 kt to 250 ft above ground level, the mean height at which the low-level or 'tactical' phase will be flown. The formation is now flying in and out of cloud below safety height in mountainous terrain and it is clear that, without INS and SKE, the mission would have to be aborted. Once at low level, the weather improves but it is a pitch-black night with no moon, and there are few identifiable features on the ground even when it can be seen. Flying at 210 kt—adjusted occasionally to maintain the required timing—all crews concentrate hard on their exacting task, navigating primarily by INS while continuing to keep station by means of SKE.

Some 20 min before the DZ, with the horizon ahead beginning to show signs of the approaching dawn, the paratroops and despatch crews begin their final preparations for the drop. About 5 min before the TAP, the stream commander orders the formation to slow down gradually to 140 kt so that the aircraft pre-dropping checks can commence. From now on, radio silence will be maintained except for

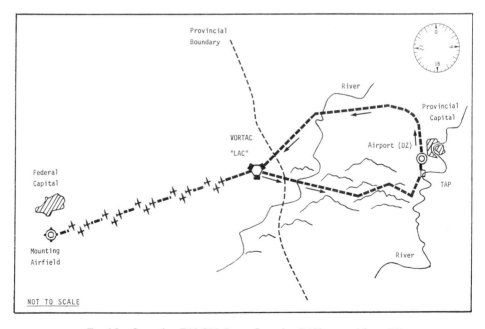

FIG 6.5. Operation TALON. Route flown by C-130s to and from DZ.

essential transmissions. As forecast, the weather in the target area is reasonably good and, with morning twilight now well established, the crews have no difficulty in pinpointing the TAP. Crossing this final check-point, the formation is 10 sec later than planned but the leader decides against an increase in speed; at this stage of the mission, it is better to allow the formation to settle down into its pre-drop configuration.

At a range of 4 nm from the DZ, with the airport hangars in sight, and with paratroops and despatch crews now at 'action stations', the formation leader reaches the predetermined 'pop-up point' where he climbs to his dropping height of 800 ft and reduces speed to 120 kt. The whole formation, now in good visual contact behind him, follows in sequence. At this higher altitude, the runway is clearly visible, providing a good reference marker for the respective personnel and platform DZs, to which the relevant aircraft now adjust their run-in. As each C-130 approaches its calculated air release point—at which its loads will be despatched—the navigator calls 'Red on!' over the intercom, followed 4 to 5 sec later by 'Green on!'. Backed up by correspondingly coloured lights which are illuminated at the despatchers' positions, these calls are respectively the warning and executive orders to the loadmaster to despatch the troops or equipment.

Having completed the drop, each aircraft descends once again to 250 ft and accelerates to 210 kt. The assault seems to have taken the rebels completely by surprise and the formation does not appear to have come under fire at any stage. Clearing the DZ area, the stream splits up as soon as possible. The nine aircraft which have dropped personnel form into three elements of three while the other six aircraft break into two elements of three. Remaining at low level and using separate

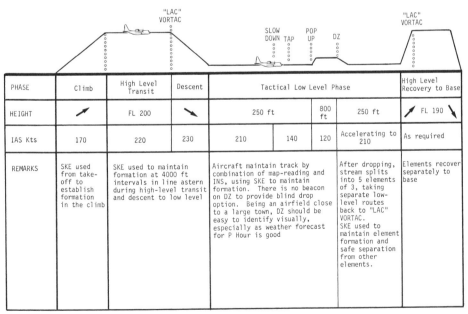

PHASE	Climb	High Level Transit	Descent	Tactical Low Level Phase			High Level Recovery to Base
HEIGHT		FL 200		250 ft	800 ft	250 ft	FL 190
IAS Kts	170	220	230	210	140	120	Accelerating to 210	As required
REMARKS	SKE used from take-off to establish formation in the climb	SKE used to maintain formation at 4000 ft intervals in line astern during high-level transit and descent to low level		Aircraft maintain track by combination of map-reading and INS, using SKE to maintain formation. There is no beacon on DZ to provide blind drop option. Being an airfield close to a large town, DZ should be easy to identify visually, especially as weather forecast for P Hour is good		After dropping, stream splits into 5 elements of 3, taking separate low-level routes back to "LAC" VORTAC. SKE used to maintain element formation and safe separation from other elements.	Elements recover separately to base

NOT TO SCALE

FIG 6.6. Operation TALON – representative flight profile.

tracks, the five elements now head back independently in the direction of the 'LAC' VORTAC, continuing to use SKE to maintain position within the element as well as to ensure safe separation from other elements. Approaching the VORTAC, each element climbs back into the airway and recovers to base.

The profile of the round-trip mission is illustrated diagrammatically in Figure 6.6.

Questions

1. List five tasks which paratroops or special forces might undertake in a limited war scenario.

2. Air-drop operations in a hostile environment can pose considerable risks to the tactical airlifters involved. How can these risks be minimized?

3. What criteria govern the selection of a TAP?

4. Give a brief explanation of the HALO technique.

5. What is the main benefit of LAPES and PLADS?

6. (a) What are the principal benefits of SKE?

   (b) What are the chief limitations of SKE?

   (c) What navigation information is provided by SKE?

7. What key items of avionic equipment do tactical airlifters need if they are to be capable of all-weather, low-level operations?

8. Define what is meant by the term 'pop-up point' in an air-drop operation.

7

Support Helicopter Operations

INTRODUCTION

Until now, we have focused exclusively on fixed-wing transport operations, emphasising their vital role in the rapid deployment of men and equipment both between and within theatres. Although the tactical operations considered thus far have normally terminated at a forward airhead or DZ, that is by no means the end of the story for either the ground or air forces involved. On the contrary, such operations are often only the prelude to further intensive activity, requiring additional deployment, redeployment and logistic support on or in the vicinity of the battlefield.

Depending on such factors as payloads, distances, terrain, and availability of airstrips, not to mention the prevailing tactical situation, some of these tasks are best performed by smaller airlifters such as the C-23A Sherpa or DHC-5 Buffalo. However, there are still many airlift tasks both in the forward and rear areas which, by virtue of its unique combination of mobility, agility and manoeuvrability, can be undertaken only by the support helicopter. The helicopter is therefore an essential complement to the fixed-wing element in the overall air-mobility and force projection equation. This Chapter will examine the principles, tasks and characteristics of helicopter transport operations before surveying some of the major aircraft types employed in this role.

As a quick glance at any recent edition of *The Military Balance*[1] will confirm, helicopters are now a significant component of many armed forces throughout the world, performing a wide variety of missions which reflect considerable differences in national doctrines and concepts. Many types are in operational service, ranging from heavily armed aircraft such as the Soviet Hind-D (a formidable combination of gunship, anti-tank platform and assault helicopter) to versions of the Sikorsky Sea King used for search and rescue duties. (Coverage of 'attack', reconnaissance and other specialist helicopters will be found elsewhere in this series, in volumes more appropriate to their particular field of employment.) This Chapter will consider what may be termed Support Helicopter Operations which themselves embrace a whole spectrum of airlift tasks from full-scale airmobile operations to routine logistic sorties by single helicopters.

TASKS

The primary role of support helicopters is to provide airlift for the tactical point-to-point movement of troops and transport of equipment, weapons and supplies. Within

PLATE 7.1 Westland Sea King (tactical transport version) carrying an underslung load. This helicopter is used in direct support of Commando amphibious assault and landing operations.

this general framework there is—as indicated above—a wide range of missions of which the most important is the provision of support for airmobile operations.

Airmobile Operations

In recent years, airmobile operations have assumed an increasingly important place in both NATO and Warsaw Pact strategy. In essence, the airmobile concept is based on the use of helicopters to provide increased mobility for ground combat forces on and around the battlefield, thereby enhancing a commander's ability to react quickly and effectively to a changing tactical situation across the entire width and depth of his sector. Controlled by the ground force commander, airmobile operations are hence an integral part of the land battle, with a vital part to play in wresting the

PLATE 7.2 Cabin of Westland Sea King (tactical transport version). Up to 30 combat-equipped troops can be carried over a radius of action of 170 nm.

initiative from the enemy while conferring greater tactical freedom and flexibility. Airmobile operations can include the following tasks:

Tactical Deployment of Troops and Weapons

Support helicopters can be used to:

- Move an attacking force from its assembly area to its start line.
- Rapidly move support weapons, ammunition and supplies.
- Move forces to attack the enemy from the flank or rear.
- Leap-frog fresh troops to maintain the momentum of an advance.
- Disengage troops for re-grouping.
- Expedite the deployment of reserves to reinforce success.

Assaults and Raids

Support helicopters can be particularly useful in assisting ground forces to undertake assaults and raids on enemy positions. Such operations take many forms. For example, it may be necessary to seize an airstrip or SAM battery, or to destroy an enemy fuel dump. Sometimes—when, for example, a bridge must be captured intact before it can be blown up by its defenders—the ground force commander may conclude that his soldiers must be landed virtually on top of the objective. This type of operation, known as a *coup de main*, aims to surprise and defeat the opposition by striking a sudden, decisive blow. However, the risks involved in a 'vertical envelopment' operation—the term used to describe a heliborne assault directly upon a defended position—are very considerable, both for the helicopters and for the troops which they are required to deliver. Ground and air commanders contemplating such an option would need to consider most carefully whether the mission would have a higher probability of success if, instead of mounting a direct heliborne assault, troops were landed nearby before completing their attack on foot. The latter tactics were adopted during the notable British raid on an Argentine outpost on Pebble Island during the Falklands war of 1982. Operating from ships under cover of darkness, troop-carrying helicopters landed special forces teams several miles from their objectives. This helped to preserve the element of surprise by preventing the not inconsiderable noise of the helicopters from being overheard by the defenders. The troops then moved overland to prosecute their attack (which resulted in the destruction of some 10 enemy aircraft) before returning to their rendezvous with the helicopters waiting to extract them.

When it is known that an intended landing zone is covered by direct enemy fire, then the positions in question should, if at all possible, be subjected to suppressive attack (either by artillery or by offensive support aircraft) immediately before the helicopter assault is due to begin. Similarly, when the landing zone is threatened by indirect fire, every effort should be made to attack the enemy weapon sites just before the assault is launched. If such pre-emptive attacks are not feasible, then the enemy positions should be subjected to immediate retaliation if they commence firing once the assault has begun. The importance of adequate firepower support underlines the need for thorough planning, close coordination and meticulous briefing of all units involved in such operations. Despite these precautions, heliborne assaults are always likely to involve a high degree of risk because so many things can go wrong. The enemy may turn out to be much stronger than expected; helicopter losses may be heavier than anticipated; supporting fire may be ineffectual; and in the confusion of battle, command and control may break down. Hence the commander who authorises an assault operation in which helicopters are tasked to land troops directly onto a defended position must first satisfy himself that the benefits of a successful outcome will hopefully outweigh the potential risks, including the probable loss of some valuable combat resources.

Infiltration

Using clandestine techniques and operating under cover of darkness, support helicopters are ideal for the insertion and recovery of patrols and reconnaissance teams, including special forces. Such personnel can be infiltrated behind enemy lines either by parachuting, abseiling, jumping from the low hover or air-landing.

Redeployment

Should the tactical situation demand, the availability of support helicopters will allow the ground commander to:

- Redeploy troops to alternative positions.
- Move support weapons quickly to threatened sectors.
- Rapidly re-group and bring his reserves into action.
- Move rearguard forces to successive delaying positions to cover a withdrawal.

Logistic Support

Helicopters can undertake a wide variety of logistic support tasks including:

- Routine daily resupply when surface or fixed-wing delivery is impossible.
- Special delivery of urgently needed weapons, ammunition or other supplies.
- Provision of airlift for demolition teams, repair personnel, route reconnaissance parties and their respective equipment.
- Movement of stocks between a fixed-wing airhead or seaport and the forward logistic base.

Cargo can either be carried internally or underslung beneath the fuselage (usually in nets) from one or more hooks. These hooks are anchored to strong-points, the main hook being positioned directly beneath the rotor head at the aircraft's centre of gravity. The choice between internal and external airlift is normally decided jointly by the air and ground unit commanders after careful consideration of the following factors:

Size and Shape of Load

- *Density*
  In general, low density loads do not 'fly' well in the underslung mode; such cargo is best carried internally if at all possible. On the other hand, externally carried high density loads (such as crates of ammunition) are less unstable than lighter cargo. This is an important characteristic since their weight may exceed the floor-loading limits, whereas the external hooks are usually stressed to carry the aircraft's maximum payload.
- *Bulk*
  Excessively bulky or awkwardly shaped loads are unlikely to fit inside a helicopter. In many cases, however, such loads can be carried externally, albeit at the expense of some restrictions on speed and range.

Nature of Load

Although wheeled vehicles can be readily loaded on and off the larger types of helicopters, non-wheeled freight can be difficult to handle unless it is palletised or containerised and the aircraft is equipped with a roller conveyor and cargo handling system. Hazardous items, such as fuel or ammunition, are usually best carried as underslung loads.

Time and Distance

- *Preparation and Loading/Unloading Times*
  Loads earmarked for airlift by helicopter must be carefully prepared and their weights accurately calculated. This takes time. External loads may take longer to prepare than internal cargo, especially if additional packaging and rigging are required. On the other hand, most helicopters can pick up and release an external load in about 2 and 1 min respectively, whereas the loading and unloading of internal cargo (depending on its type and quantity) may take much longer. In operational situations where time is of the essence, the net balance of these factors may well decide the loading configuration to be adopted.
- *Radius of Action*
  In other circumstances, the distance over which a load is to be airlifted may be the overriding factor. Drag and instability induced by underslung loads combine to lower a helicopter's transit speed, increase its fuel consumption over a given distance and hence reduce its effective range. Cargo must therefore be carried internally if the mission requires the aircraft to operate over its maximum radius of action. Conversely, as the required radius of action decreases, other factors such as loading and unloading times become more significant, shifting the balance in favour of internal loads.

Cargo Handling

Although underslung loads usually require more rigging and lashing equipment than internal cargo, they can often be delivered to the exact position where they are needed, even to the extent of ammunition being lowered directly onto a flat-bed truck or beside a battery of guns. On the other hand, helicopters carrying heavy internal loads not only require landing areas which are relatively firm, smooth and level, but also tend to need special handling equipment when non-wheeled freight is being transloaded. Dependence on such equipment, which is unlikely to be available in forward areas, can be a serious disadvantage.

Freedom of Manoeuvre

As already indicated, helicopters carrying external loads must fly at reduced speeds. Moreover, the inherent instability of large underslung loads may make an aircraft more difficult to handle, particularly at night, in cloud or other conditions of poor visibility. For obvious reasons, aircraft in this configuration must also fly at higher altitudes, thus further restricting their tactical freedom of manoeuvre *vis-à-vis* helicopters carrying internal loads, which are usually able to employ 'nap-of-the-earth' tactics[2] to achieve maximum security and surprise. That said, a helicopter flying at low level with underslung cargo may be able to withstand an attack or indeed a serious malfunction if the external load can be immediately jettisoned. Using

PLATE 7.3   RAF Puma disengaging from a gun which it has carried by cargo hook
to a forward location.

a quick-release device provided for this purpose, the pilot can instantly increase
manoeuvrability while reducing the all-up weight by an amount which may be crucial
to his aircraft's survival. This ability to shed payload instantaneously is, of course,
denied to the pilot of a helicopter carrying internal cargo.

Casualty Evacuation

Expeditious evacuation of battlefield casualties improves their chances of survival,
eases the burden on commanders in the forward areas and does much to sustain
morale. Helicopters are ideally suited to this important task, especially when surface
access is rendered difficult by terrain or enemy action. Casualties may either be
emplaned in the conventional way or, if the aircraft is unable to land, winched aboard
using techniques similar to those employed on search and rescue operations. Patients
are normally carried inside the cabin but, in extremis, can be underslung in litters
on a special harness beneath the fuselage. The latter is an emergency procedure which
would be used only as a last resort and then only over short distances. In minor
conflicts involving small numbers of casualties, it may be feasible to dedicate some
helicopters to the casualty-evacuation role, but in larger-scale scenarios, a commander
may face an awkward dilemma in weighing the need for such missions against high-
priority operational tasks. In these circumstances, casualty-evacuation may have to

be conducted on an opportunity basis, airlifting injured personnel back to the rear areas on aircraft which had brought in troops, equipment or supplies.

Reconnaissance and Observation

Although the ground commander may have no option but to conserve his dedicated support helicopters for their primary roles (airmobile operations and logistic airlift) such assets, if available, can be used to good effect for reconnaissance, surveillance and observation. Typical reconnaisance missions include the survey of lines of communication and key points, area surveys to identify potential sites for field HQs and artillery sites, and assessments of battle damage at airfields attacked by the enemy. Observation missions, on the other hand, are usually intended to keep a specific area under surveillance in order to detect, track and report enemy movements. By virtue of their ability to cover relatively large areas in short timescales and to operate in terrain which may be inaccessible to ground reconnaissance vehicles, most helicopters are ideal for such tasks.[3]

Assistance in Command, Control and Communications

On today's fast-moving battlefield, a ground commander needs to be mobile. One way of enhancing his mobility is to establish a heliborne command post, an option which is virtually essential to the effective control of an airmobile operation. In some armed forces, a few helicopters (modified to provide additional communications facilities) are specifically earmarked for this role. More often than not, however, helicopters are pressed into this type of service on an *ad hoc* basis, with the commander and his staff having to make do with whatever communications happen to be available. The range, payload and (in most cases) all-weather capability of medium-sized support helicopters make them particularly suitable for use as airborne command posts but, as in the case of reconnaissance, the ground commander may be unable to justify their employment on such tasks, which are more likely to be assigned instead to smaller aircraft such as the UK Army Air Corps Gazelle or Lynx. Similarly, smaller helicopters are more likely to be used than the larger logistic types to carry commanders between HQs and subordinate units, or to distribute important messages during periods of electronic silence or in the event of a break-down in radio communications.

Internal Security Operations

Not all military activity takes place on or adjacent to the battlefield or within the context of direct confrontation with a readily identifiable enemy. Sometimes a state will find that the most immediate threat comes not from beyond its borders but from within, in the shape of groups who are prepared to use force to de-stabilise or replace the existing government or political system. When this happens, the civilian police and gendarmerie (if any) may be unable to maintain internal security without considerable help from the armed forces. This has been the case for many years in Northern Ireland, where substantial British forces have been used to assist the local security forces in containing the terrorist activities of the 'Irish Republican Army'.

In this long-running operation, as in many others around the world, the support helicopter has proved invaluable in helping the security forces to tackle this most difficult and sensitive of military tasks. The many missions which support helicopters can undertake in an internal security operation include:

- Rapid deployment of security forces in response to incidents.
- Positioning and extraction of rooftop and surface observation teams.
- Deployment and recovery of patrols, including special forces.
- Establishment of road blocks to isolate suspected terrorists.
- Rapid deployment of bomb disposal teams.
- Assist in establishment of cordons during sweep and search operations.
- Aerial reconnaissance and surveillance, using electronic sensors and other specialised equipment.
- Reinforcement and resupply of garrisons and police stations which, for security reasons, are inaccessible by road.
- Dispersing or monitoring crowds. Helicopters can give useful warning of assemblies and can keep watch on gatherings and marching mobs.
- Rapid casualty evacuation.
- Movement of VIPs.

Humanitarian Tasks

Just as fixed-wing airlifters may be employed in aid of the civilian authorities, so support helicopters can be called upon to perform a wide range of non-military tasks. Many states have helicopters which are dedicated to the search and rescue role but when such aircraft are unavailable or insufficient in number, support helicopters can often be quickly switched to undertake this type of mission. It is in the field of humanitarian operations, however, involving natural catastrophes such as earthquakes, floods, famine and blizzards that the inherent versatility of the typical support helicopter comes into its own. In these situations, rotary-wing airlifters can be invaluable in supplying food, evacuating casualties or providing mobility for rescue teams and equipment.

CHARACTERISTICS AND LIMITATIONS OF SUPPORT HELICOPTERS

While support helicopters have many valuable characteristics and applications, they are also subject to a number of restrictions and limitations. The more important of these characteristics and limitations—many of which are common to all rotary-wing aircraft used in land operations, including armed and attack helicopters—are outlined below.

Characteristics

Versatility

While most helicopters can undertake a variety of roles, many tend to be optimised for particular types of mission such as anti-armour operations. Support helicopters,

PLATE 7.4   Soviet Mi-10 Harke, sometimes known as the 'flying crane'. Its quadricycle landing gear and under-fuselage clearance of 12.3 ft allows the aircraft to straddle (and if necessary taxi over) loads before they are secured for external airlift. These special capabilities make the Mi-10 ideal for many emergency civilian tasks, including the positioning of prefabricated huts and shelters.

however, are exceptionally versatile—as indicated by the span of tasks outlined earlier in this Chapter.

Mobility

Support helicopters are able to operate in most areas of the world irrespective of geographical features such as water, swamps, forests or glaciers. Furthermore, thanks to their ability to use confined landing sites requiring little or no preparation, they normally enjoy substantial freedom of manoeuvre within a particular operational theatre. Even when they cannot land, support helicopters can often load and unload personnel and underslung loads.

Flexibility

Coupled with the versatility and mobility described above, the helicopter's ability to switch rapidly from one task or location to another affords the ground commander a very useful degree of flexibility. For example, it can be used to deploy troops and support weapons over distances and within timescales that would otherwise be impossible. Moreover, unlike fixed-wing airlifters, support helicopters can usually land very near the positions from and to which its payloads are to be respectively collected and delivered.

Speed

One of the support helicopter's chief advantages is its relatively high speed compared with that of competing surface transportation systems. This is an important attribute because—as we have already emphasised—the rapidity with which a deployment can be completed may well determine the outcome of an operation. That said, there could be some situations where surface movement may be almost as quick if not faster than helilift. For example, over shorter distances with reasonably good tracks or roads, trucks may take less time to assemble, load and complete a given logistic task. Over longer distances and difficult terrain, however, the helicopter's significantly higher transit speed will usually be decisive.

Agility and Manoeuvrability

The helicopter's agility and manoeuvrability, and its ability to operate at very low altitudes (using nap-of-the-earth techniques to avoid detection) collectively mean that it can often be used to achieve surprise. On the other hand, helicopter operations may be compromised if radar reflections are detected by low-level air defence radar, or if cockpit/fuselage glint or rotor blade flicker are spotted by hostile forces. Surprise may also be lost due to engine or rotor noise in circumstances where other noise levels are low, although it is not alway easy to identify a helicopter's position or track solely from the noise which it is emitting.

Limitations

Vulnerability

The vulnerability of helicopters to hostile action in forward operational areas is sometimes a contentious issue. However, many informed analysts would agree that it is relatively difficult to acquire and bring weapons to bear on a helicopter which is being skilfully flown close to the ground, below the tree line (if any) and making full use of available cover. On the other hand, a helicopter is probably at its most vulnerable when flying between 50 and 2,000 ft above ground level, when it presents an attractive target to slow-flying enemy combat aircraft, attack helicopters, surface-to-air missiles (SAM) and small arms. Even so, helicopters have on the whole proved remarkably resilient and able to absorb a surprising degree of battle damage without catastrophic consequences:

> 'The war in Vietnam provided the United States with a situation and terrain which were well suited to helicopter operations. Medium and heavy lift helicopters and the ubiquitous utility helicopter all played important parts in this war, although it was, perhaps, the coming of age of the helicopter gunship which was of most interest. For those who maintained that the helicopter was too vulnerable to survive on a modern battlefield, the statistics proved interesting. Very few helicopters were lost as a result of enemy anti-aircraft fire, and this against a

proliferation of small arms and heavy MGs and also against sophisticated AA missiles. The US Army had no doubt that the helicopter had proved its worth as a battlefield weapon.'[4]

Whether helicopters will retain this degree of survivability in the future is another question. As far as can be judged from open sources, Soviet helicopters operating in Afghanistan seem to have faced a sterner challenge than the US helicopters in Vietnam. While precise statistics are not available, it is known that a number of Soviet helicopters have been shot down by Stinger SAM fired by mujahideen guerrillas. Similarly, a number of Soviet helicopters involved in the civil war in Angola have been lost both to SAM and to offensive fixed-wing aircraft of the South African Air Force.

Setting aside such factors as differences in terrain and weather patterns (which may have had some bearing on the different loss rates in Vietnam and Afghanistan) it seems that advances in technology and tactics will probably determine the balance of helicopter survivability in the future. Meanwhile, it is likely that SAM, whether radar-controlled or visually laid, will continue to pose the greatest threat to support helicopters, despite the availability of passive and active countermeasures such as warning sensors, jammers, chaff and infra-red decoys. Other measures which help to enhance survivability include extensive use of armour plating to protect the crew and certain of the more critical components, and duplication of vital systems to provide some degree of redundancy. Operations at night, when helicopters become a notoriously difficult visual target, may also help although these tactics may be offset by the problems inherent in missions mounted under cover of darkness. (See section on night operations below.)

On balance, experience gained in the many 'limited' wars of recent years confirms that support helicopter operations are entirely feasible in forward areas provided that they are conducted with due regard for the prevailing threat, exploiting the aircraft's agility and manoeuvrability, and employing an appropriate mix of protective devices. Nonetheless, if over-exposed to known or suspected anti-helicopter weapon systems (especially SAM) the loss rate could become unacceptably severe. In the final analysis, it is for the commander to weigh the risks against the potential gains of such exposure.

Performance

Helicopter performance is affected by a combination of the following factors:

- Engine power and rotor lift decrease as air density reduces with increases in altitude and temperature. Operations at high altitudes and temperatures may necessitate significant reductions in payload and manoeuvrability.
- Load-carrying capability varies widely according to helicopter type and the amount of fuel in tanks. As with fixed-wing airlifters, maximum payload cannot normally be carried over the maximum range but only over shorter distances. As the distance to be flown increases, payload will usually have to be traded for fuel.
- As discussed earlier in this Chapter, it is often necessary or more efficient for a helicopter to carry payload underslung from hooks beneath the fuselage. However,

this configuration has several drawbacks. For example, underslung loads reduce manoeuvrability and hinder terrain-screening profiles, as well as requiring the aircraft to be flown well below its normal transit speed. Moreover, the extra drag induced by external loads must be countered by additional power, thereby increasing the fuel burn and hence reducing the range.

Night operations

While there are tactical advantages in operating under cover of darkness, night missions can also pose considerable problems for support helicopters as a result of the difficulties entailed in map reading, flying close to the ground and using unlit landing sites without adequate visual references.

However, the advent of highly accurate and reliable INS, as well as night vision goggles (NVG), means that these problems can now be largely overcome. Such aids are extremely expensive, not merely in terms of the goggles themselves but also due to the need to provide a compatible cockpit. Without NVG, however, helicopters operating at night may have to accept the following limitations:

- Higher transit altitudes.
- Reduced terrain screening.

PLATE 7.5 Litton M-925/M-927 Night Vision System in flip-up position. Night vision goggles are sensitive to the red and near infra-red wavebands, which is that part of the spectrum where most of the energy in starlight and moonlight is found and where there is good reflectivity from the earth's surface. Using the latest types of NVG, an experienced pilot can safely fly nap-of-the-earth missions in moonlight or starlight.

- Slower speeds.
- Increased separation between aircraft in both time and space.
- Additional preparation, control and illumination of landing sites. The personnel used to perform these tasks would have to be inserted on foot, by parachute or by pathfinder helicopter, any of which could compromise the security of the operation.

Weather Restrictions

Depending on its severity and duration, bad weather can seriously disrupt helicopter operations. For example, strong surface winds can make it very hazardous to start and stop rotors and, in extreme cases, may even halt operations until they have abated. As far as *en route* weather is concerned, most modern helicopters can be flown on instruments (ie, without reference to the ground) though pilots must do so at a safe height and be able to descend to their landing sites either visually or using electronic guidance. Support helicopters may sometimes be able to turn poor visibility to their advantage, but while it is reasonable to expect that such conditions will hinder visual acquisition by the enemy, the aircraft will still be liable to detection by electronic and thermal sensors. Moreover, poor weather may cause serious navigation difficulties for aircraft without sophisticated aids. Overall, however, the unique characteristics of helicopters mean that they can usually operate at low level in less visibility and under a lower cloud base than fixed-wing airlifters.

Ground Security

Support helicopters are not merely vulnerable to hostile action when airborne. Whilst on the ground, they also present an attractive target for attack by enemy air and ground forces and must be protected accordingly. It is sometimes difficult for helicopter units to organise an effective defence against a ground threat since they are not usually provided with the necessary resources, in terms of properly trained and equipped personnel, to undertake this specialist task. Hence their operating bases should preferably be sited in assembly areas or other locations where they can take advantage of the presence of other units which do have adequate defences. The proximity of friendly forces in the base area may also afford some degree of protection from air attacks, either from local surface-to-air weapons or from an air defence umbrella established to protect the area as a whole. Regardless of whether such systems are available, helicopters can also counter the air threat by dispersal, concealment (eg, in farm buildings or warehouses) or camouflage. However, such measures are not always easy to implement. For example, dispersal is all very well but can pose difficulties for aircraft maintenance as well as compounding the problem of providing adequate ground defence. Moreover, helicopters are not easy to manhandle and camouflage even when the environment is favourable; on rugged or exposed terrain, they may be virtually impossible to conceal. In such circumstances, commanders may either have to redeploy the helicopters to a better site or accept a high degree of risk that they will be detected and attacked.

Logistic Support

- *Fuel*

  Like most rotary-wing aircraft, support helicopters have a high fuel consumption. This characteristic not only reduces their radius of action as providers of airmobility and logistic support for other units, but also poses logistic problems for the support helicopters themselves when operating away from their main bases. For example, the establishment and replenishment of adequate fuel stocks at the required locations are sometimes tasks which can be performed only by helicopters, thus absorbing valuable airlift which will not then be available for operational missions.

- *Maintenance and Availability*

  Being relatively complicated, helicopters usually need considerable routine maintenance by technicians who accompany the aircraft whenever they deploy into the field. If and when a support helicopter unit is subsequently redeployed, a proportion of its airlift capacity will usually need to be allocated to move these personnel, together with their special servicing equipment, to the new location. Though exacting by comparison with most fixed-wing aircraft, the average helicopter's maintenance schedule is sufficiently flexible to allow an increased rate of operations for limited periods—for example, to exploit a spell of favourable weather, or in response to tactical demands. However, a price must be paid for flying at intensive rates, in terms of subsequent reduced availability whilst any outstanding essential maintenance is completed.

Deployment

Despite their relatively limited range, most support helicopters can self-deploy over considerable distances, either by making frequent stops for refuelling or by using specially fitted ferry fuel tanks to permit longer intervals between staging airfields. Some US helicopters, including examples of the CH-47D Chinook and CH-53E Super Stallion, are equipped for in-flight refuelling from C-130 tankers (see Plate 7.6) but this technique is not widely used.

Even if AAR is used, the support helicopter's low cruising speed and inability to climb above adverse weather encountered *en route* mean that such self-deployment can be painfully slow. For example, it would normally take at least 3 days for a detachment of Pumas to fly from the UK to Eastern Turkey, a journey of just over 2,000 nm. As an alternative to self-deployment, some types of support helicopter can be airlifted into or within a theatre by transport aircraft such as the C-5, C-141 or C-130, but this can be an expensive process in terms of both time and labour if the helicopters in question have to be substantially dismantled before loading and then reassembled at the destination. Ideally, support helicopters and large fixed-wing airlifters should be designed with each other in mind, so that the airlifters can carry the helicopters into forward airheads with little or no dismantling required beyond folding the main rotors. As will be seen in Chapter 8, the importance of this concept has been recognised by the manufacturers of the C-17 airlifter which has been designed to carry, *inter alia*, a load of four UH-60A utility helicopters in the rotor-folded configuration. This capability represents an important step along the road towards integrating the support helicopter into the overall airlift system.

PLATE 7.6   Two Sikorsky CH-53E Super Stallions of the US Marine Corps refuelling
in flight from a KC-130T operated by the Marine Corps Reserve.

GEOGRAPHICAL LIMITATIONS

In addition to the general restrictions described above, support helicopter operations
are subject to a further series of regional restrictions which vary with terrain and
climate as described below.

Jungle Operations

In dense jungles, such as those of Indo-China or Central America, where surface
routes are few and far between, helicopters offer an excellent means of deploying,
resupplying and extracting ground forces. For example, an infantry patrol can save
many days of arduous marching by being airlifted into and out of its operational
area. On the other hand, there are limits to the number of helicopter landing sites
which can be provided in such an environment and, in these circumstances, military
operations tend to be based on an established series of clearings which will soon
become known to an enemy. Other problems likely to be encountered include:

- Reduced performance and payload associated with high ambient temperatures.
- Severe thunderstorms which can hamper maintenance and make flying conditions
  very difficult.
- Difficulty of navigation over areas of uniform appearance with few landmarks
  and unreliable maps.

Cold Weather Operations

Although support helicopters offer valuable mobility to a land commander in cold
weather environments such as North Norway, Greenland or South Georgia, the
conditions typically encountered in these areas pose serious problems including:

- Extremely low temperatures which may cause oils and other aircraft fluids to congeal.
- Severe storms, blizzards and strong winds which can bring helicopter operations to a standstill.
- Snow blown up by the rotors, obscuring visibility on take-off and landing.
- Featureless terrain and unreliable maps, making navigation difficult.
- Difficulty of concealment.
- Need for additional logistic support, such as heated maintenance shelters and de-icing equipment.

Desert Operations

Support helicopter operations in desert regions may be hampered by:

- Very high temperatures which degrade performance and hence reduce payload.
- Sand blown up by the rotors, obscuring visibility on take-off and landing, and causing engine damage by sand ingestion unless special filters are fitted.
- Featureless terrain and unreliable maps, making navigation difficult.
- Difficulty of concealment.
- Difficulty of maintenance in conditions which can vary from extreme heat by day to intense cold by night.

Attempt to Rescue Hostages from Iran, 1980
At this point, it is worth digressing to recall that sand ingestion was a major factor in the failure of the attempt in April 1980 to rescue 53 hostages who were being held by the Iranians in the US Embassy in Teheran. Units of all four US Services (including special forces) participated in this operation, with airlift and logistic support provided by six C-130s and eight Sikorsky RH-53D helicopters. The eight helicopters took off from the aircraft carrier USS *Nimitz* for a flight of about 500 nm to Desert One (an airstrip in the Iranian desert some 260 nm south-east of Teheran) where they were to rendezvous with the C-130s. Soon after crossing the Iranian coast, at least three RH-53Ds were caught in a sandstorm, forcing one aircraft to turn back to the *Nimitz* with serious technical problems. Ironically, as events were to turn out, this aircraft was carrying the hydraulic maintenance equipment for the entire RH-53D force assigned to the operation. Soon afterwards, one of the remaining helicopters suffered a major hydraulic failure and force-landed in the desert, its crew being picked up by one of the other aircraft. On arrival at Desert One, a second RH-53D developed a major hydraulic failure which could not be repaired since the necessary equipment was now back on the *Nimitz*. This reduced the number of serviceable helicopters to only five and since the planners had stipulated that the next phase of the operation could not proceed without at least six helicopters, the mission was aborted at this stage. Unfortunately, worse was to come. While preparing (in pitch darkness) to refuel from one of the C-130 tankers prior to redeploying back to the *Nimitz*, one of the helicopters collided with one of the troop-carrying C-130s, killing eight crewmen and destroying both aircraft in the subsequent fire. All of the remaining personnel then withdrew by C-130,

the four operational RH-53Ds being abandoned in case they had been damaged by the blast caused by the explosion as the two aircraft collided. This ill-fated operation, involving the loss of eight lives, seven helicopters and one C-130 provides an exceptionally graphic illustration of the difficulties which can arise when using helicopters for long-range desert operations.

Operations in Mountainous Terrain

Support helicopters may be required to operate in mountainous country either in temperate latitudes or in any of the three scenarios described above. Although their performance declines with altitude, helicopters can perform many useful tasks in such terrain including the movement of troops and supplies within much shorter timescales than would be possible on foot or in vehicles using mountain tracks. They can also move pickets to otherwise inaccessible locations, maintain isolated garrisons which might be too hazardous to support by surface convoy, and establish forces on commanding positions without great physical effort by the troops concerned, thus ensuring that they arrive fresh and ready for action. Nevertheless, helicopter operations in mountainous terrain are subject to certain restrictions. For example, helicopters cannot land on steep slopes or boulder-strewn sites. When it is necessary to deplane troops or supplies in such circumstances, this may have to be done from the low hover, or personnel may have to descend by rope. This technique was sometimes used by Soviet units in Afghanistan.

PLANNING SEQUENCE

The basic principle in the planning of support helicopter operations is that the air plan must always be based upon and subordinate to the ground tactical plan, not vice versa. In practice, this means that (as with parachute assault operations) the planning process should be undertaken in reverse order, beginning with the ground force's mission and working backwards to the preparatory and mounting phases. Thus a typical plan involving an assault by heliborne infantry should develop in the following sequence.

Ground Tactical Plan

This covers all aspects of the ground forces' operations once they have landed. It includes such essential ingredients as the mission objectives, landing sites to be used, command and control, timings, tactics to be employed and intelligence on enemy dispositions. It should also specify what fire support, if any, is required and available either from artillery, naval gunfire, fixed-wing combat aircraft or attack helicopters. This tactical plan is produced by the ground force commander who usually has overall responsibility for the operation.

Air Plan

The air plan is normally produced by the officer who commands the support helicopter force, in conjunction with the commander who is responsible for providing and coordinating any air-delivered firepower that is required by the operation. Designed to complement the ground plan as effectively as possible, the air plan should cover all facets of air participation, including any missions flown by attack helicopters or fixed-wing combat aircraft as well as those helicopters tasked to carry troops and equipment. The plan should include types and numbers of aircraft involved; timings and routes; heights, speeds and formations to be flown; pick-up and drop points; and arrangements for command, control and communications. If close air support aircraft are involved in the operation, it may be necessary to provide a heliborne forward air control team to designate their targets, in which case the pertinent details must also be covered in this plan.

Loading Plan

Devised by the ground commander's staff, the loading plan should ensure that troops and weapons are allocated to helicopters in accordance with the requirements of the ground tactical plan. Particular care must be taken to ensure that key personnel and items of equipment are distributed across more than one aircraft, to abate the effect of any losses whilst *en route* to the landing site.

SUPPORT HELICOPTERS IN CURRENT OPERATIONAL SERVICE

The remaining section of this Chapter will focus on just a few of the many types of support helicopter currently in military service around the world.

Mi-17 Hip H

The Mikhail Mil Mi-17 (code-named Hip H by NATO), which made its first appearance in the West at the Paris Air Show of 1981, is in effect an updated and more powerful version of the earlier Mi-8 Hip. Although the Mi-8 has now been operational for over 20 years, large numbers remain in front-line service with the Warsaw Pact, as well as many other armies and air forces. The fact that so many Mi-8s are still operated by the USSR is not simply another example of that country's reluctance to discard aircraft which can still offer useful service, but also a reflection of the aircraft's continuing importance as the backbone of the Soviet logistic helicopter fleet. To date, some 10,000 Mi-8s and Mi-17s have been built in all—an impressive production run by any standards, which is likely to increase much further before manufacture of the Mi-17 is eventually phased out. At present, a total of about 1,800 Mi-8s and Mi-17s are thought to be in service with the Soviet armed forces, the two types frequently operating alongside one another.

Superficially the Mi-17 bears a strong resemblance to the Mi-8, but its tail rotor has been relocated on the port side of the vertical stabiliser and its twin engines have shorter nacelles than those on the Mi-8. Both models have the same cabin configuration and can carry similar loads, but the Mi-17 is primarily employed as a cargo-lifter while the Mi-8 is more often used to carry troops. However, the main difference is the Mi-17's superior performance, thanks to its more powerful TV3-117MT engines, each developing 1,900 eshp compared with the 1,500 eshp produced by the Mi-8's TV2-117A engines. This extra power allows the Mi-17 to carry about

PLATE 7.7   Mi-8 Hip.

a ton of payload more than the Mi-8; it can also fly higher, faster and further. Moreover, the Mi-17's engines are linked so that, if one loses power, the other automatically compensates. If one engine fails completely, the other automatically produces maximum power at the emergency rating of 2,200 eshp. In order to enhance the Mi-17's ability to operate from unprepared sites in deserts and elsewhere, deflectors can be fitted to the air intakes to reduce the ingestion of sand and other harmful material into the engines.

In the passenger configuration, the Mi-17 can carry 32 fully equipped combat troops (five more than the Mi-8) over a radius of 130 nm. Alternatively, the helicopter can be readily converted to accommodate 12 stretchers and one medical attendant. Troops normally emplane and deplane through a sliding door on the port side of the cabin (see Plate 7.8) but they can also use the rear entrance where folding clamshell doors permit full-width access to the cargo compartment. Hook-on ramps can be attached to the rear sill of the cabin to facilitate the loading and unloading of small military vehicles, whilst an integral winch and pulley system is provided to facilitate the loading and unloading of heavy or bulky equipment. The external cargo hook is stressed to carry loads weighing up to 6,614 lb.

Despite being primarily employed in the logistic and airmobility roles, many Mi-17s and Mi-8s are equipped with weapon pylons mounted on outriggers at each side of the aircraft. This allows the helicopter to carry 128 × 57-mm rockets in four pods (the usual choice of weapons when armament is deemed essential) or up to four anti-armour missiles. Provision for such weapons confers a degree of self-protection or defence-suppression capability which could prove valuable during opposed operations, but the Mi-8's size, weight and lack of agility strongly militate against its employment in an offensive support role.

PLATE 7.8 Soviet troops deplaning from Mi-17.

Summary of Leading Particulars

Crew: Two pilots.

External dimensions:	Main rotor diameter	69.9 ft (21.29 m)
	Length (excluding tail rotor)	60.5 ft (18.42 m)
	Height (to top of main rotor head)	15.6 ft (4.76 m).
Internal dimensions:	Cargo compartment length	17.5 ft (5.34 m)
	Cargo compartment width	7.7 ft (2.34 m)
	Cargo compartment height	5.9 ft (1.80 m)
	Cargo compartment volume	812 cu ft (23 m³).

Engines: Two Isotov TV3-117MT turboshafts, each delivering 1,900 eshp.

Basic weight: 15,653 lb (7,100 kg).

Maximum payload:	Internal	8,820 lb (4,000 kg).
	External	6,614 lb (3,000 kg).

Troop capacity: 32.

Maximum fuel (standard tanks): 3,200 lb (1,450 kg).

Normal take-off weight: 24,470 lb (11,100 kg).

Maximum take-off weight: 28,660 lb (13,000 kg).

Range with maximum fuel: 260 nm.

Typical cruise speed: 130 kt.

Mi-26 Halo

Four years before the AN-124 Condor, the world's largest fixed-wing airlifter, made its Western debut at the 1985 Paris Air Show, the USSR had used this same aeronautical shop window to unveil the Mikhail Mi-26, the heaviest helicopter built to date and the first to employ an 8-bladed main rotor. In 1982, the year after its appearance at Paris, the Mi-26 (code-named Halo by NATO) established a number of impressive weight-lifting records including an ascent to 15,092 ft with a payload of 44,090 lb (20,000 kg).

Some idea of this helicopter's impressive size and cargo capacity can be gleaned from Table 7.1 which compares certain vital statistics of the Mi-26 with those of the C-130H.

The Mi-26's ability to lift such heavy and bulky loads says much for the skill of the designers who have made judicious use of advanced materials, including titanium, to achieve a power-to-weight ratio which enables the aircraft to become airborne at a total all-up weight which is twice its basic empty weight (see Summary of Leading Particulars). A further example of the Mi-26's versatility is its ability to carry its

PLATE 7.9   Mi-26 Halo.

TABLE 7.1 *Relative dimensions of C-130H/Mi-26 cargo holds and comparison of maximum payloads.*

	C-130H	Mi-26
Cargo hold		
Length	40.1 ft	39.4 ft
Width	10.2 ft	10.7 ft
Height	9.1 ft	10.4 ft
Maximum payload	45,000 lb	44,090 lb

maximum payload of 44,090 lb either internally or externally, a closed-circuit television system being provided to allow the crew to monitor underslung loads during all stages of flight.

The cargo hold, which is equipped with a winch and twin electric hoists on an overhead rail, can accommodate various permutations of light armoured vehicles, trucks, artillery and palletised supplies. Alternatively, the helicopter can carry 84 combat-equipped soldiers, using 20 tip-up seats along each wall, a further 40 seats specially installed for the trooping role and four seats in a small, separate cabin behind the flight deck. There are three passenger doors in the cabin, one at the front on the port side and one on each side aft of the main landing gear. All of these doors are downward-hinged and fitted with integral steps. Vehicles and other cargo can be loaded and unloaded through the rear of the aircraft where a pair of clamshell upper doors, and a downward-hinged lower door with integral folding ramp, open to afford full-width access to the hold. This arrangement is illustrated in Plate 7.10. As a further aid to loading and unloading, the length of the main landing gear struts can be hydraulically adjusted during operations from uneven surfaces.

PLATE 7.10   Mi-26 on the ground, with rear doors open.

Overall, the Mi-26 is a highly effective airlifter which combines the great advantage (common to all support helicopters) of being able to operate from any reasonable landing site, with the ability to carry payloads which compare favourably in terms of size and weight with many fixed-wing transports. The comparison is less flattering, naturally, in terms of range and speed. Whereas the C-130 can carry its maximum payload (45,000 lb) at 300 kt over a range of 2100 nm, the Mi-26 can only carry its maximum load (44,090 lb) at 137 kt for a fraction of this distance. Nevertheless, the Mi-26 is a remarkable aircraft which is certain to enhance the general mobility of the Soviet armed forces.

Summary of Leading Particulars

Crew: Two pilots (aircraft commander in left-hand seat), navigator, flight engineer—all on flight deck. Loadmaster in cargo hold.

External dimensions:	Main rotor diameter	105.0 ft (32.00 m)
	Fuselage length (excluding tail rotor)	110.7 ft (33.73 m)
	Height (to top of main rotor head)	26.7 ft (8.15 m).
Internal dimensions:	Cargo compartment length	39.4 ft (12.00 m)
	Cargo compartment width	10.7 ft (3.25 m)
	Cargo compartment height	10.4 ft (3.17 m).

Engines: Two Lotarev D-136 turboshafts, each producing 11,400 eshp, with air intakes designed to prevent ingestion of foreign matter. If one engine fails, power on the remaining unit is automatically increased to the emergency maximum rating.

Basic weight: 62,170 lb (28,200 kg).

Maximum payload: 44,090 lb (20,000 kg).

Normal take-off weight: 109,125 lb (49,500 kg).

Maximum take-off weight: 123,450 lb (56,000 kg).

Range with maximum fuel (at maximum take-off weight)**:** 430 nm.

Typical cruise speed: 137 kt.

UH-60A Black Hawk

Conceived and developed as a replacement for the UH-1 Iroquois utility helicopter, the Sikorsky UH-60A Black Hawk first flew in 1974. Four years later, it began its productive life with the US Army—by far the largest operator of these aircraft with over 800 (of 930 ordered) now in front-line service.

The UH-60A is a highly versatile, multi-purpose helicopter. In its primary role as a battlefield air mobility vehicle, it can carry a section of 14 fully equipped troops or a 105-mm howitzer on the cargo hook plus the gun's five-man crew and 50 rounds of ammunition in the cabin. Alternatively, it can be used to evacuate casualties, haul equipment and supplies (inside the cabin or underslung), undertake certain reconnaissance tasks and act as an airborne command post. Moreover, the UH-60A is itself highly air-portable. For example, one Black Hawk can be carried in a C-130, two in a C-141B and six in a C-5B, using special transportation rigs to minimise preparation at the point of departure and reassembly at the destination. Hence these helicopters can be rapidly deployed over strategic distances and commence operations

PLATE 7.11   US Army Black Hawk carrying 16 Hellfire missiles on External Stores
Support System (ESSS).

in distant theatres within the same timescales as the ground forces they are assigned
to support.

If further versatility is required, the UH-60A can be fitted with an 'external stores
support system' (ESSS) which provides four removable pylons (two on each side of
the fuselage) on which ancillary fuel tanks, weapons or various specialist pods can
be carried. The ESSS is stressed to carry 5,000 lb on each side, allowing 1,703-litre
and 870-litre fuel tanks to be mounted respectively on each of the inboard and
outboard pylons. This quantity of additional fuel allows the aircraft to self-deploy
over sectors of 1,200 nm without refuelling. In battlefield missions, or on special
operations where some measure of self-defence capability is considered prudent, the
ESSS can be used to carry such weapons as guns, rockets, Hellfire anti-armour
missiles and/or ECM pods. However, the UH-60A is not an attack helicopter and a
sensible balance must therefore be struck between the carriage of armament or extra
fuel, and the airlift of troops or equipment, if the aircraft's capacity to carry the latter
is not to be seriously degraded. That said, the ESSS makes the UH-60A particularly
suitable for long-range, clandestine missions behind enemy lines with small teams of
special forces. The Black Hawk's capacity for such tasks has been recently enhanced
by the development of exterior and interior lighting systems which are compatible
with NVG, thereby allowing the aircraft to be flown low-level at night into unlit
landing sites.

Summary of Leading Particulars

Crew: Two pilots in cockpit, plus crewman in cabin.

External dimensions:

Main rotor diameter	53.7 ft (16.36 m)
Length (excluding refuelling probe)	50.1 ft (15.26 m)
Length (air-portable configuration, ie rotors and tail pylon folded)	41.3 ft (12.60 m)
Height (to top of main rotor head)	12.3 ft (3.76 m)
Height (air portable configuration)	8.8 ft (2.67 m).

Cabin capacity: 385 cu ft (10.9 m³).

Engines: Two General Electric T700-GE-700 turboshafts, each delivering 1,560 eshp.

Basic weight: 10,624 lb (4,819 kg).

PLATE 7.12   US Army UH-60A carrying external load.

Maximum payload: Internal   5,636 lb (2,556 kg)
                              External (cargo hook)   8,000 lb (3,630 kg)
                              External (ESSS)   10,000 lb (4,536 kg).

Troop capacity: 14 combat-equipped troops (in high density configuration). Eight seats can be replaced by four stretchers, or removed to make space available for cargo.

Fuel capacity (internal tanks): 1,363 l (approximately 2,250 lb).

Normal take-off weight: 16,260 lb (7,375 kg).

Maximum take-off weight: 20,250 lb (9,185 kg).

Range with maximum internal fuel (at maximum take-off weight): 320 nm.

Range with maximum internal fuel (plus fuel carried on ESSS):

Two × 870-l tanks   880 nm
Two × 870-l tanks plus two × 1,703-l tanks   1,200 nm.

Typical cruise speed: 145 kt.

CH-47D Chinook

The Boeing Vertol CH-47D is the latest version of the highly successful series of Chinook medium transport helicopters which first entered service as the CH-47A in 1962, followed by the CH-47B in 1967 and the CH-47C in 1968. The major customer for these aircraft was the US Army which took delivery of 354 CH-47As, 108 CH-47Bs and 270 CH-47Cs, and which has since placed a contract with Boeing Vertol for the conversion of 436 of these aircraft to the recently introduced D-model standard under a programme not scheduled for completion until 1993.

The CH-47D, which began its operational life in 1984, bears little resemblance to earlier variants of the Chinook beyond sharing the same silhouette; for example, its payload is nearly twice that of the CH-47A. Not surprisingly, the upgrading of A, B and C models to this new and much improved standard entails substantial modification. After stripping the aircraft down to its basic structure and replacing any defective components, the Boeing engineers virtually rebuild it from scratch, incorporating a whole range of improvements including more powerful engines, improved transmission, new glass fibre rotor blades, an advanced flight control system to reduce pilot workload, triple cargo hooks and NVG-compatible cockpit, as well as improved avionic, electrical and hydraulic systems. The net result is a versatile, up-to-date helicopter with better performance, reliability and maintainability than its predecessors, and a further life of at least 20 years.

Capable of flying in most weather conditions, the CH-47D is a flexible airlifter which can undertake a variety of aeromedical, rescue or parachuting tasks in addition to its primary function as a vehicle for the rapid movement of troops and cargo. Operating at night with INS, NVG and various protection devices (such as radar warning sensors, IR jammers and chaff dispensers) fitted to enhance its survivability in a hostile environment, the Chinook can also be used for long-range special forces missions, its cabin being sufficiently spacious to accommodate an extra fuel tank as well as some 20 soldiers and their equipment. For everyday operations, it can carry 44 troops (in 33 sidewall and 11 centre-aisle seats) but more than twice this number can be carried by allowing soldiers to sit on the floor when operational necessity dictates. In the aeromedical role, the cabin can accommodate 24 stretcher patients plus two medical attendants.

PLATE 7.13   US Army CH-47D lifting a bulldozer weighing nearly 25,000 lb. This bulldozer is an important item of Army field equipment, being used for a wide variety of combat-related construction activities.

Many of the CH-47D's impressive qualities are directly related to its tandem rotor system. Whereas torque created by single-rotor helicopters must be counteracted by a tail rotor, sometimes to the detriment of performance, the Chinook's counter-rotating rotors cancel out each other's torque. More importantly, the tandem rotor system improves handling precision in the hover and 'broadens' the centre of gravity thereby allowing greater flexibility in the disposition of both external and internal loads. This means, for example, that exceptionally long items of cargo can, if necessary, be carried with the ramp open. The tandem rotor design also permits the entire length of the fuselage to be utilised for payload, with no space wasted on an empty tailboom. Added to the fact that fuel is carried in blister tanks running along each side of the lower fuselage, this design provides a cabin with a constant cross-section, as illustrated in Figure 7.1.

PLATE 7.14   CH-47D carrying a 155-mm howitzer.

For such a compact helicopter, the CH-47D (which at 51 ft is only a foot longer than the UH-60A Black Hawk) has a deceptively spacious cargo compartment which can accept various combinations of heavy or bulky freight, including two jeep-sized vehicles (tracked or wheeled). Using the built-in winch, such loads can be quickly loaded via the full-width rear ramp which can either be fully lowered to the ground or adjusted to truck-bed level as required. The quadricycle landing gear (with double-wheel units at the front and single wheels at the rear) provides valuable stability

PLATE 7.15   Troops deplaning through rear door of US Army CH-47D.

during loading and unloading, enabling the aircraft to operate from slopes of up to 20°. Skis can be fitted to prevent sinking in snow, mud or marshland and, if absolutely necessary, the helicopter can even operate from water in conditions up to sea state 3, thanks to its unique amphibious capability.

As already noted the CH-47D has three hooks for external loads. The centre hook can carry 26,000 lb while the forward and aft hooks are each rated at 17,000 lb. The ability to lift such heavy loads externally is operationally advantageous because urgently needed equipment and weapons, such as bulldozers and howitzers, can be carried intact with no time needed for dismantling before flight or reassembly at the delivery point. Another important benefit of the triple hook system is that it allows outsized cargo to be carried in a more stable configuration—and hence at significantly greater transit speeds—than if a single hook is used. For instance, the Chinook can normally fly at more than 115 kt when carrying a large load secured to all three hooks, whereas it is limited to about 50 kt when flying with the same type of load on only one hook. A further advantage of the triple hook system is that it allows separate external loads (such as fuel containers or ammunition pallets) to be delivered to three separate locations during a single sortie. Alternatively, one CH-47D can lift a complete artillery element with gun, vehicle and ammunition respectively underslung from the three hooks while the gun crew is carried inside the cabin. That said, techniques do vary from operator to operator. The RAF, for example, prefers its triple hook-equipped Chinook HC1s to uplift a heavy gun with the centre hook only, as this eases the centre of gravity problem. US CH-47Ds, on the other hand, normally carry these weapons underslung from the front and rear hooks in order to achieve better stability.

By any objective standards, the CH-47D has been particularly well designed for its key role as a provider of airmobility and logistic support for ground forces on and around the battlefield. Its ability to operate at intensive rates for sustained periods in the field has much to do with its high degree of built-in self-sufficiency, as exemplified by its:

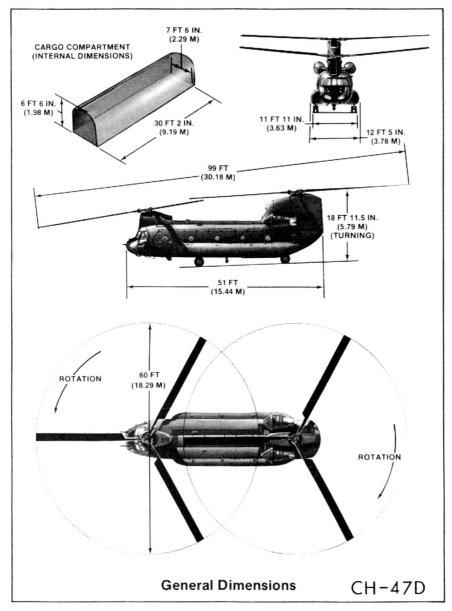

FIG 7.1   General dimensions of CH-47D.

- Auxiliary power unit which drives all electrical and hydraulic systems.
- Widespread application of modular technology, simplifying repairs and routine maintenance.
- On-board panel to perform 26 self-monitoring and inspection functions covering the hydraulics, transmission and engine systems.

Finally, the CH-47D can be flown on instruments in most weather conditions,

PLATE 7.16   Tracked vehicle inside CH-47D.

including light icing and moderate turbulence. It can also operate in all but the most extreme climates, being designed to tolerate temperatures ranging from $-54°$ to $+52°C$.

Summary of Leading Particulars

Crew: Two pilots and two crewmen. In RAF Chinook squadrons, the co-pilot may be replaced by a navigator.

Dimensions: See Figure 7.1.

Rotor system: Two three-bladed counter-rotating rotors, driven by interconnecting shafts which enable both rotors to be powered by either engine.

Engines: Two Avco Lycoming T55-L-712 turboshafts, each delivering 3,750 eshp (emergency rating 4,500 eshp).

Basic weight: 22,452 lb (10,184 kg).

Maximum payload: Internal load (100-nm radius)   18,000 lb (8,164 kg)
External load (30-nm radius)   20,700 lb (9,389 kg).

Troop capacity: 44 combat-equipped soldiers or 24 stretchers plus two attendants.

PLATE 7.17   Canadian Armed Forces CH-47D being loaded with miscellaneous cargo.

PLATE 7.18   RAF Chinook carrying separate loads on three hooks.

Fuel capacity: 3,899 l (approximately 6,500 lb) in self-sealing tanks contained in external fairings.

Maximum take-off weight: 50,000 lb (22,680 kg).

Typical cruise speed: 140 kt with internal load (less with underslung loads, speed varying with configuration).

PLATE 7.19   US Army truck being secured for airlift using dual hooks.

PLATE 7.20   UH-60A (minus rotors) being carried on single hook of CH-47D.

Questions

1.   Describe the basic concept of airmobile operations.

2.   List four types of airmobile operations.

3. What factors influence the choice between internal and external airlift of cargo?

4. Give five examples of missions which support helicopters can undertake in an internal security operation.

5. What measures can be taken to enhance the survivability of support helicopters?

6. What are the disadvantages of underslung loads?

7. List five problems which may hamper support helicopter operations in cold weather.

8. What is the basic principle which governs the planning of support helicopter operations?

9. How does the performance of the Mi-17 compare with that of the Mi-8?

10. What are the main advantages of the CH-47D's tandem rotor system?

8

The Future

AIRLIFT'S ROLE IN FUTURE FORCE PROJECTION

As the preceding chapters have confirmed, air transport forces make a substantial contribution to the exertion of political and military power, performing a wide spectrum of missions with often impressive results. This role is likely to remain closely linked to the parent state's broader politico-military responsibilities and aspirations, which in turn will continue to be conditioned largely by the prevailing international climate.

Many parts of the world (notably the Middle East, the Gulf, Southern Africa and Central America) are beset by chronic conflicts which, though relatively limited in scale and scope at present, contribute to a pattern of Third World instability which will probably persist well into the next century. Compared with areas such as these, Europe—despite its sharp ideological division into Western and Eastern blocs—is virtually a paragon of stability, tension between NATO and the Warsaw Pact having been eased first by the 1986 Stockholm Agreement and then by the 1987 US/Soviet treaty to eliminate a whole class of intermediate nuclear missiles by 1990. However, while generally welcomed as a stepping-stone to possible reductions in strategic nuclear systems, this treaty has also served to accentuate the importance of conventional forces to NATO's deterrent posture. Given the Warsaw Pact's superiority in in-place conventional capability, NATO's requirement to be able to ferry substantial US reinforcements across the Atlantic (and to redeploy its forces within Europe) as rapidly as possible will become even more critical than hitherto, thereby placing additional emphasis on the availability of an appropriate mix of inter- and intra-theatre airlift assets.

Such resources are likely to remain equally indispensable to the prosecution of foreign and domestic policy beyond the NATO/Warsaw Pact area, where speed of reaction may make all the difference to the success or otherwise of any military intervention in pursuance or defence of national interests, including the policing of international trade routes and maintenance of access to vital raw materials such as oil. In short, with the international scene certain to remain unpredictable and volatile due to political friction, regional wars and economic uncertainties, the need for armed forces to enjoy a high degree of mobility will probably increase rather than diminish.

Hence it seems reasonable to suppose that air transport operations will assume an even greater significance during the 1990s than in the preceding decade, and not just in those states such as the USA and USSR which have leading roles to play on the world stage. Air mobility is also likely to become increasingly important in Third World countries where:

- Threats may emanate from more than one direction.

182

- Geography and lack of an appropriate infrastructure inhibits rapid surface movement.
- Large standing armies are not only an economic burden but also a potentially de-stabilising influence.

In the West, airlift forces are likely to become increasingly significant for quite different reasons. In addition to the geo-strategic imperatives mentioned earlier, dependence on air mobility will probably grow as the democracies come under more and more pressure to reduce the size of their armed forces in response to a combination of political, economic and demographic factors. If and when such reductions are implemented, the ability to deploy these smaller forces rapidly to wherever they are needed will be even more crucial than before to the defence posture of the states concerned.

After identifying some of the problems which must be overcome if airlift forces are to cope with these growing responsibilities, this Chapter will take a detailed look at several new aircraft that have been specifically designed to meet the challenges which can be anticipated up to and beyond the turn of the century.

CURRENT PROBLEMS

Despite their developing role as key instruments of national defence, many air transport forces are currently hampered by a variety of shortcomings which seriously restrict their capacity to undertake their full range of commitments. Even the USAF's Military Airlift Command (MAC), widely acknowledged as the most powerful and effective airlift force in the world today, faces most if not all of the problems outlined below.

Shortfall In Total Capacity

At present, few air transport forces have sufficient capacity to meet either their peacetime or probable wartime tasks. This is not simply a function of the universal military adage which dictates that demand for airlift will always exceed supply, but reflects real shortfalls in capability. For example, it has been calculated that MAC currently has only some two-thirds of the airlift capacity it needs to support US forces already deployed around the world, with an actual inter-theatre capacity of 48m ton-miles per day compared with a requirement to carry 66m ton-miles per day. The figures themselves are less important than their operational implications which add up to the stark fact that, as things stand, US commanders in the field would probably not receive, within the required timescales, the amount of equipment and supplies they would need to wage an effective war outside the USA. This situation exists even though MAC has in recent years obtained a substantial injection of resources including the upgrade of all Starlifters to C-141B standard, the expansion of the C-5 fleet to a total of 127 aircraft, the acquisition of 60 KC-10s and enhancements to the CRAF programme.[1]

Shortfall in Outsize Capacity

An especially significant feature of the overall shortfall in airlift is the lack of capacity to carry the very largest items of military equipment. All military transports can carry bulk cargo in the form of pallets, containers or miscellaneous freight and most, including the AN-12, C-130 and C-160, can carry 'oversize' loads such as small trucks and trailers, medium artillery pieces and the smaller types of helicopter (suitably dismantled to fit into the hold). However, only aircraft such as the AN-22, AN-124 and C-5 can carry 'outsize' loads like heavy trucks (some of which weigh up to 40,000 lb), self-propelled artillery, armoured fighting vehicles, CH-47Ds and other medium-sized helicopters, combat engineer equipment (including bulldozers and bridgelaying equipment) and main battle tanks (weighing in some cases more than 120,000 lb). Depending on the scenario, threat and mission, the ratio of outsize loads will probably vary between 30 and 40% of the total tonnage to be deployed, but the actual percentage is less important than the fact that it is mainly this component which gives the force its 'teeth', the smaller items consisting mainly of support equipment and supplies. To put this problem into perspective, it is of course true that realism in the formulation of foreign policy objectives and defence budgets means that few states apart from the USA and USSR need ever contemplate the airlift of such exceptionally large loads. Nevertheless, the two Superpowers' present inability to airlift more than a modest proportion of their outsize combat equipment has an important bearing on their ability to intervene rapidly in overseas crises, and hence has wider implications for regional and global stability. On the other hand, the fact that neither Superpower has unlimited capacity to become embroiled in local conflicts might be considered no bad thing!

Shortage of Reception Airfields

From a Western perspective, the relative shortage of airfields capable of accepting the larger airlifters is especially worrying. An official US study has shown that, while there are some 10,000 runways of at least 3,000 ft × 90 ft throughout the free world (excluding the USA), only about 850 of these are suitable for C-5 and C-141B operations (See Figure 8.1).

The problem would be particularly acute in the event of a major reinforcement of the Central Region of NATO. Although the Federal Republic of Germany has approximately 140 airfields with runways of at least 3,000 ft, only 18 of these have sufficient ramp space for C-5 and C-141B operations. Moreover, over a third of that total ramp space is located at one airport (Frankfurt) which, with so many potential eggs in one basket and being only 80 miles from the Inner German Border, would presumably be high on the list of Warsaw Pact targets. On the other hand, most of the other large airfields in Western Europe are so far from probable operational areas that payloads on strategic aircraft would have to be moved forward over considerable distances, either by surface or by air after transloading onto smaller aircraft such as the C-130. (The drawbacks of trans-shipment are covered separately below.) As a result, there would be added congestion not only on the surface access routes between the operational zone and the major reception airfields but also at the airfields themselves. Moreover, the fact that relatively few large airheads would be

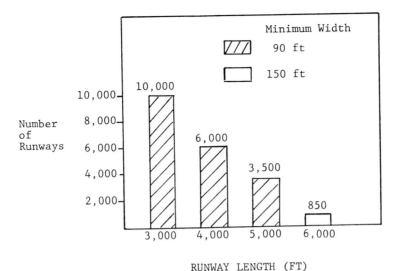

FIG 8.1. Number of runways in free world (excluding USA).

available greatly simplifies the enemy's task in disrupting or interdicting airlift operations, while reducing the options available to NATO planners and commanders.

Inefficiency of Transloading

The fact that airlifters such as the C-5, C-141B, KC-10, Tristar, VC-10 and B-707 cannot operate into smaller airstrips close to the operational zones means that their payloads must be off-loaded at major airfields for onward delivery to the ultimate destination. Some of this cargo can be airlifted forward by C-130, C-160 or similar aircraft but these are unable to carry outsize loads which must therefore complete their journey by surface transport.

Trans-shipment is an inefficient way of doing business. It is time-consuming, expensive and entails duplications of cargo handling equipment as well as additional personnel for operations, aerial port and maintenance teams. However, there is often no alternative at present, owing to performance limitations which preclude the current generation of large airlifters from delivering their loads directly into forward airheads.

Age of Airlift Fleets

Many NATO and Warsaw Pact airlifters are now ageing, some of them having been in front-line service since the 1960s. However, the productive life of a transport aircraft is not so much based on its chronological age as on the number of flying hours which the relevant authorities have decided it can safely accumulate without exceeding certain stipulated fatigue parameters. In other words, an airlifter's life is measured by its time spent in the air rather than by the dates on a calendar. Compared with commercial airliners, military transports consume their fatigue life at a relatively slow pace, especially if—as in the case of the C-5A and C-141A—it is

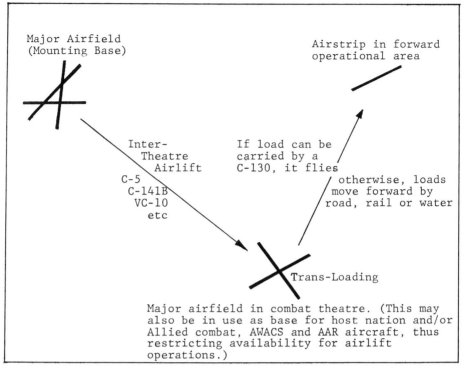

FIG 8.2.   Trans-shipment concept.

extended by mid-life updates or major modification programmes. This is why many of today's military airlifters are expected to remain in operational service to the turn of the century and beyond, despite having been designed 30 years ago. Unfortunately, whilst aircraft can be re-engined, strengthened, stretched, re-sparred, modified for AAR and equipped with the latest avionics, little can be done to improve their basic design and performance. An airlifter designed in the 1950s for missions anticipated in the 1960s and 1970s will almost inevitably prove less than ideal for the 1980s, much less the 1990s. For example the C-141B, despite its excellent qualities in other roles, will never be able to operate from short airfields. In short, the longevity of military transports is a mixed blessing. On the one hand, it allows the authorities to achieve the maximum return on the considerable financial investment but on the other, it means that commanders must accept penalties in the later stages of the aircraft's service which can seriously constrain their concepts of operations.

Other Problems

Obsolete designs are also at the root of other problems common to many of the military transports currently in service. Although some aircraft are better equipped than others, there is a general need for greater all-round reliability and for systems which require as little maintenance as possible, especially when operating away from the home base. Moreover, relatively few airlifters are protected with equipment such as radar-warning sensors, missile jammers and chaff dispensers. Against a background of more urgent priorities elsewhere for their hard-pressed defence budgets, few nations seem prepared to fit such devices to their ageing transport fleets. This could be false economy; the modest sums involved are out of all proportion to the wider flexibility conferred, and probable savings in crews, aircraft and loads, during airlift operations where there is a known SAM threat.

Framework for Future Concept of Operations

Owing to wide variations in the defence policies of individual states, it is impossible to develop a universal concept of air transport operations except in the broadest terms. Nevertheless, it is worth noting a number of basic principles which are now emerging to influence the future philosophy of airlift forces and which seem certain to govern the design and employment of the next generation of medium and large transport aircraft.

As many states increase the proportion of their ground forces which can be rapidly deployed by air, in line with a general trend towards greater mobility, the overall demand for airlift will inevitably rise. Furthermore, operating within as well as between theatres, air transport forces must in future expect to be required to deliver all items of air-portable equipment directly into most 3,000 ft airfields. This will ease congestion at the major airfields; overcome the inefficiencies of transloading; and most important of all, allow loads to be delivered as quickly and as close as possible to where they are needed. Adoption of this new concept, which will significantly enhance operational efficiency and flexibility, poses a formidable challenge not only for the designers of the new aircraft required to make it work, but also for the defence budgets which will have to finance these expensive new assets.

FUTURE AIRCRAFT

C-17

In August 1981, the USAF announced that McDonnell Douglas had won a keenly contested competition with Lockheed and Boeing to design and build an advanced airlifter for service with MAC in the 1990s and beyond. After several years of further debate and uncertainty, caused mainly by Congressional misgivings about the new aircraft's cost and viability, McDonnell Douglas finally received a full-scale development contract in December 1985. Production of a prototype is now under way and the first flight is planned for mid-1990. Designated the C-17, the new airlifter is expected to enter operational service with MAC in 1992. The USAF has placed an initial order for 210 C-17s at a fixed price of $91m per aircraft at 1986 prices.

In launching the original competition, the USAF took the unusual but shrewd step of defining the new aircraft's operational mission in terms of representative scenarios and timescales rather than specifying its characteristics. This ensured that the design was driven from the outset by such important factors as:

- Size, weight and amount of combat equipment to be carried.

- Inter-theatre distances.
- Locations of airfields likely to be used.
- Airfield dimensions (including taxiways and parking ramps as well as runways).
- Ground manoeuvrability.
- Reliability.

Take-Off Performance and Range

The C-17 is intended to be a multi-purpose airlifter, capable of intra- as well as inter-theatre missions. However, it seems probable that it will be mainly employed in the latter role, initially alongside and eventually as a replacement for the C-141B which in turn may displace the older C-130s in the Air Force Reserve and Air National Guard. Based on representative scenarios and data on mounting airfields (including airfield lengths, elevations and average temperatures) the USAF stipulated that the new airlifter should be able to carry its maximum payload from a runway of not more than 8,000 ft over a minimum distance of 2,400 nm. The manufacturers predict that the C-17 will not only meet this 'bottom line' requirement but also be able to carry a substantial payload over ranges up to 3,500 nm. This is illustrated in the graph in Figure 8.3 which compares the C-17's performance in terms of range and payload with that of existing MAC aircraft. This projection indicates that the C-17

PLATE 8.1. Artist's impression of C-17 operations at forward airstrip. (McDonnell Douglas).

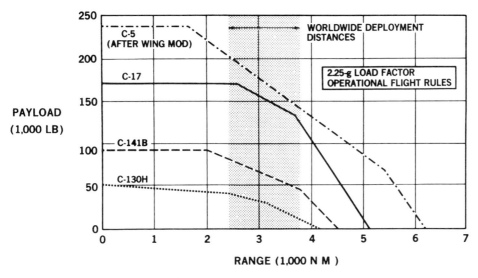

FIG 8.3. Comparison of C-17 range and payload with existing MAC aircraft.

should be able to carry some 90% of the C-5A's payload over sectors of 2,500–3,500 nm. Such a capability (if realised) is in no small measure attributable to the provision of a particularly good engine, the Pratt & Whitney PW2037 turbofan. Already well proven in airline service with the B-757, the thrust and fuel efficiency of this engine are expected to enable the C-17 to more than achieve the take-off and range criteria set by the USAF.

Landing Performance

Another key feature of the C-17's proposed concept of operations is that it must be able to carry outsize loads such as main battle tanks into austere forward airheads with runways of only 3,000 to 4,000 ft. In order to meet this exacting requirement the aircraft will need, *inter alia*, to have low take-off and landing speeds, relatively short take-off and landing runs, and be capable of touching down at the required point with great precision. The C-17 will achieve these characteristics by using 'powered lift' whereby engine efflux is directed onto and through large double-slotted flaps to increase the coefficient of lift on the supercritical wings. (See Figure 8.4.) As

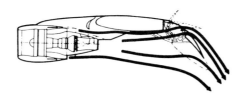

C-17
POWERED LIFT PROVIDES
SHORT-FIELD CAPABILITY

FIG 8.4. Powered lift.

a result, the C-17 will be able to fly at much lower speeds than would otherwise be possible for such a large and heavy aircraft.

Powered lift will also enhance the C-17's landing performance by enabling it to descend into a destination airfield on a 5° glide-path. This is considerably steeper than the 3° approach normally flown by transport aircraft. A steep glide-path, combined with twin head-up displays to assist the pilots in flying a very accurate approach and achieving an extremely accurate touch-down, is expected to reduce the air-run component of the landing manoeuvre by a significant margin. Test data collected from earlier trials with the McDonnell Douglas YC-15 (an experimental aircraft used to confirm the feasibility of the powered lift concept) suggests that the C-17 should be able to achieve touch-down within 250 ft either side of the desired position, thus further reducing the landing distance required. The air-run allowance from a conventional 3° approach is much longer (see Figure 8.5).

A third important element in the C-17's landing performance is its revolutionary thrust-reverser system which can be selected in flight immediately before touch-down. Existing reversers discharge engine efflux forward, downward and to the sides, stirring up dust and other loose material that is so often a feature of unpaved surfaces. The sudden restoration of forward power towards the end of the landing run amid such a cloud of debris may cause it to be ingested into the engines where it can cause considerable damage. However, in another example of innovative technology, the C-17's engines have been designed to deflect reverse efflux in forward and upward vectors (as illustrated in Figure 8.8) thereby allowing reverse thrust to be selected at any speed on any surface without the risk of ingesting harmful debris.

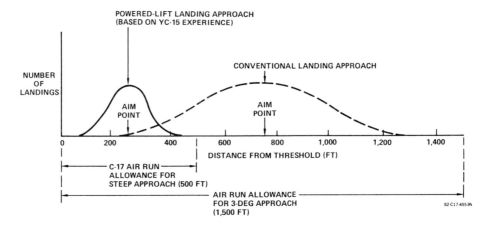

FIG 8.5.   Powered lift reduces runway length required for landing.

Ground Manoeuvrability

As indicated in Chapter 2, one of the most important parameters in planning and mounting an airlift operation is the capacity of the reception airfield's parking ramp. By restricting the type and number of aircraft that can be simultaneously on the ground, it is this factor together with the turn-round time (which varies with aircraft type) that governs the maximum flow and hence the build-up rate of the troops and equipment being delivered. This is especially significant at small, austere airstrips which tend to have narrow runways and taxiways as well as rather confined ramps. With that in mind, the USAF was careful to specify that the new airlifter should be able to operate into airfields as indicated in the top half of Figure 8.6, which means that it must be able to complete a 180° turn within a runway width of only 90 ft. Additionally, at airfields with parallel taxiways or turnaround loops at the ends of the runway, the C-17 must be able to use runways with a width of only 60 ft.

Scenarios used as the framework for the mission requirements dictated that two C-17s must be capable of operating from the smaller of the ramps (75,000 sq ft) while three C-17s must be able to park simultaneously on the larger ramp (120,000 sq ft). These dimensions, which effectively governed the new aircraft's maximum wing span and fuselage length, make an interesting comparison with those required for C-5 operations (see lower section of Figure 8.6). The area of 500,000 sq ft needed to accommodate two C-5s would be sufficient for nine C-17s, thanks to the C-17's smaller vital statistics and superior manoeuvrability (see Figure 8.7). The C-17's manoeuvrability is derived from two key factors. First, it will be able to accomplish

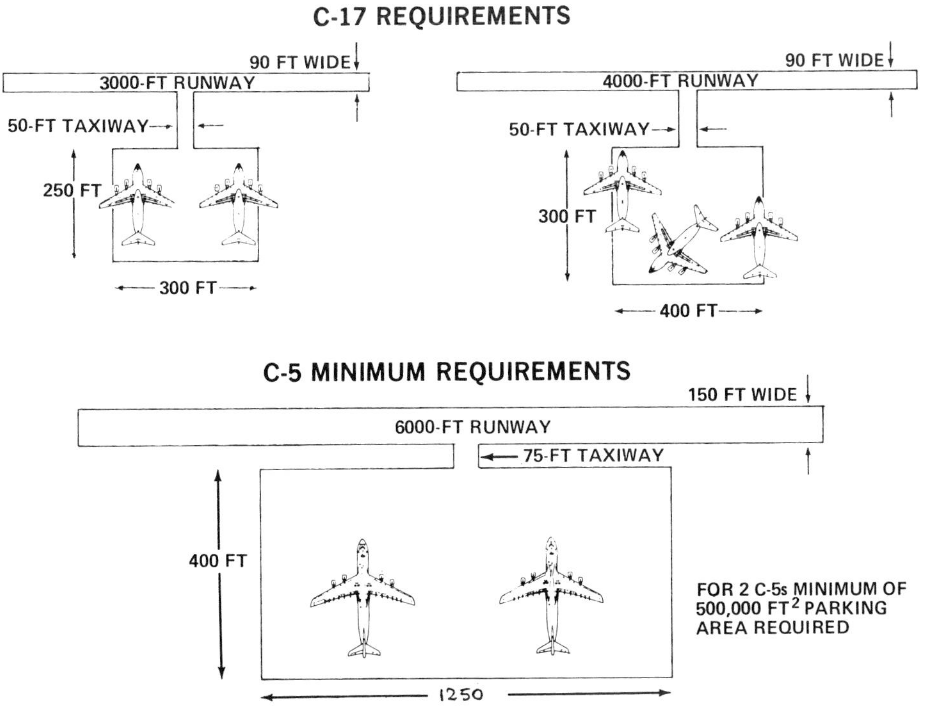

FIG 8.6. Airfield requirements for C-17 and C-5 operations.

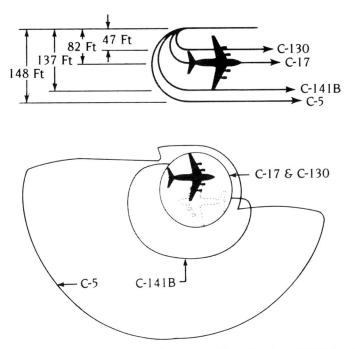

FIG 8.7. Comparison of C-17's ground manoeuvrability with existing MAC aircraft.

a 180° turn in only 82 ft (an impressive achievement for an aircraft with a wing span of 165 ft). Second, in line with yet another USAF specification, it will be able to reverse up a 2% incline with maximum payload and fuel for 1,000 nm.

Although the C-17's payload over a sector of 3,000 nm is expected to be some 25,000 lb less than that of the C-5, its quicker turn-round time (about an hour compared with the two to three hours required for a typical C-5 turn-round) and ability to squeeze into smaller spaces means that a C-17 operation would—in situations where parking areas are restricted—achieve a much higher cargo through-put than a C-5 airlift over a given period.

Engine-Running Offloads

As explained earlier, engine-running off-loads (EROs) are sometimes used to reduce turn-round times and thereby increase the flow of missions through an airhead. In other situations—where, for example, the reception airfield is under threat—EROs assume added significance as a means of minimising periods of aircraft vulnerability. Despite their operational value, however, EROs with existing airlifters can create hazardous conditions for aerial port and maintenance personnel working around the aircraft during the turn-round. This is because engine efflux (even at 'ground idle' power) is strong enough to generate dangerously hot exhaust gases and a backwash which may stir up clouds of dust and flying debris. This problem has been overcome on the C-17 by equipping it with a revolutionary thrust-reverser system which enables EROs to be conducted with the engines selected to 'reverse idle' power. A key feature

PLATE 8.2.   Artist's impression of C-17s air-dropping containers.

of this system is that reverse efflux is deflected upward at an angle of about 30° so that there is no hazard from either engine wash or airborne debris either in front or aft of the aircraft. The auxiliary power unit (APU), which can itself generate dangerously high levels of blast, heat and noise, has also been designed with ground operations in mind. By placing it at the apex of the rear fuselage (see Figure 8.8) the

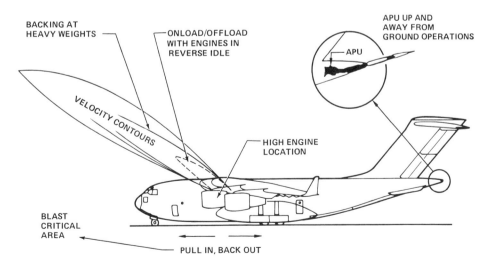

FIG 8.8.   C-17 ground operations using upward-vectored reverse thrust.

designers have ensured that the C-17's APU can be operated without hazard to any personnel working in its vicinity.

Projected Payloads

The more rectangular the shape of an aircraft's cargo hold, the greater the extent to which its volumetric capacity can be exploited. With this principle in mind, McDonnell Douglas have ensured that virtually every cubic foot of the C-17's cavernous hold can be utilised, an important feature if the aircraft's weight-lifting capabilities are not to be offset by limitations on bulk. Another key factor in achieving maximum utilisation of available space is the ability to carry cargo on the ramp. Whereas the existing family of MAC airlifters can only accept relatively light ramp loadings, the C-17—with a ramp built to the same load-bearing specifications as the floor of the cargo hold—can carry up to 40,000 lb, which is almost as much as the maximum payload of the C-130. These differences in ramp capability are illustrated in Figure 8.9 which compares the dimensions of the C-17's cargo hold with those of the C-130, C-141B and C-5.

In weight-lifting terms, the C-17's payload capacity has been maximised by the extensive use of titanium and advanced composite structures to lighten its basic weight. This has helped McDonnell Douglas to project a maximum payload of 172,200 lb, nearly twice that of the C-141B. Added to the provision of a spacious hold specially designed to accommodate large trucks and other oversize military vehicles in double rows, this means that the C-17 can carry not only twice as much weight but also twice as much equipment as the C-141B. The net result is that one C-17 should be able to do the work of two C-141Bs, thus increasing the airlift flow-rate whilst eliminating the need for trans-shipment by delivering payloads directly into forward airfields. Designed to operate with only one loadmaster, the C-17 will

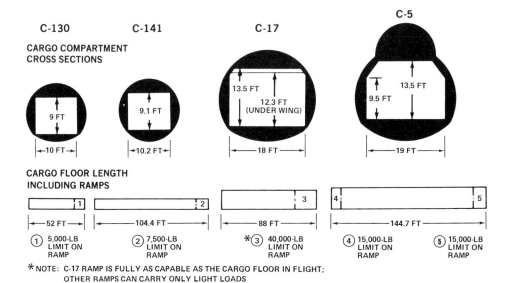

FIG 8.9. Comparison of MAC aircraft cargo holds.

be able to carry 18 standard pallets or equivalent containerised loads on its built-in roller-conveyor system. The roller-conveyor can be rapidly dismantled to provide a flat-floor configuration for various permutations of combat vehicles and helicopters. Some indication of the C-17's cargo-carrying versatility is given in Figure 8.10.

Air-Drop Capability

The C-17's tactical flexibility will be further enhanced by its ability to air-drop small teams of special forces, up to 102 paratroops, or many types of combat equipment and supplies. Taking advantage of its full-width ramp and single rear door, it will also be able to use the LAPES technique (described in Chapter 6) to deliver loads of up to 60,000 lb from a height of only 10 ft above the DZ. Indeed, the C-17 will be the only aircraft in the MAC inventory capable of delivering outsize loads either by LAPES or by using conventional air-drop techniques. This is illustrated in Figure 8.11.

Summary of Design Features

Although the innovative C-17 concept represents the integration of several state-of-the-art aeronautical technologies, all of these have been fully validated by flight-testing or operational application on previous aircraft. Allied to an exhaustive programme of wind tunnel and other tests on C-17 mock-ups and models (see Plates 8.3 and 8.4) this means that both manufacturer and customer can be reasonable confident that the advanced performance forecast for this aircraft will actually be realised. The more important design features of the C-17 are shown in Figure 8.12, which also summarises the benefits of the various technologies involved.

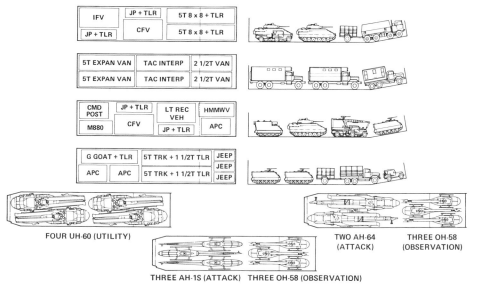

FIG 8.10.   Examples of side-by-side loads carried by C-17.

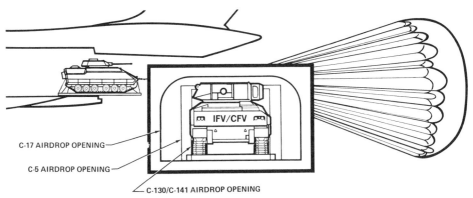

FIG 8.11. Dimensions of US infantry/combat fighting vehicles in relation to rear exits of MAC aircraft. Only the C-17 can air-drop such outsize loads.

Summary of Leading Particulars

Crew: Two pilots and one loadmaster.
External dimensions: Length   175.2 ft (53.4 m)
                          Wingspan   165.0 ft (50.3 m)
                          Height   55.1 ft (16.8 m).
Internal dimensions: See Figure 8.9.
Engines: Four PW2037 turbofans each producing 37,000 lb of static thrust.
Maximum payload: 172,200 lb (78,108 kg).
Maximum take-off weight: 570,000 lb (258,546 kg).
Seating capacity:
Sidewall (permanently installed)   54 (27 on each side of fuselage)
Centreline (stored on board)   48
Palletised (10 seats per pallet)   100 (10 pallets)
Maximum seats   154 (100 on pallets plus 54 on sidewall seats).
Aeromedical capacity: 48 stretcher patients, 102 sitting patients.
Range with maximum payload: 2,400 nm.
Cruise speed: 0.77 Mach at 28,000 ft.

OTHER PROJECTS

The USA is not the only state, nor McDonnell Douglas the only aerospace company, to be currently investing time and money in future airlift systems. Against the background of the changing operational imperatives and present-day inadequacies described earlier, all major air forces are now trying to grapple with the requirement to replace large numbers of their existing airlifters in the late 1990s or early part of the next century. While little is known of Soviet initiatives, it is safe to assume that the *VTA* will be pressing just as hard as MAC for resources to fund a major re-equipment programme over the next 10 to 15 years.

Although the C-17 promises to be an extremely effective and versatile aircraft, it is by no means the total answer to MAC's current shortfall in capability. To date, the USAF has been authorised to purchase only 210 of these advanced airlifters.

PLATE 8.3. One-tenth scale model of C-17 being prepared for tests in radio frequency anechoic chamber.

This is due partly to the huge cost ($91m per aircraft at 1986 prices) and partly to the aircraft's *raison d'être* as the replacement for the C-141B. Accordingly, the C-17 is primarily destined for the inter-theatre role where it should excel at delivering outsize and oversize loads directly into forward airfields. Despite its capacity for in-theatre operations, the C-17's employment on such missions—even if it could be spared from its strategic duties—would not normally represent the best use of such

PLATE 8.4. One-twentieth scale model of C-17 being prepared for wind tunnel tests
of structural loads with landing gear down and aft cargo door open.

an expensive asset. In any case, despite its impressive short-field performance, the C-17 does not have the STOL capabilities that will probably be an essential feature of the C-130's replacement. In short, the C-17 acquisition is only one element in MAC's overall modernisation programme; additionally, it needs a new and dedicated tactical transport not merely to replace its large fleet of C-130s but also to increase its overall intra-theatre capability. It was with that in mind that the Pentagon recently awarded

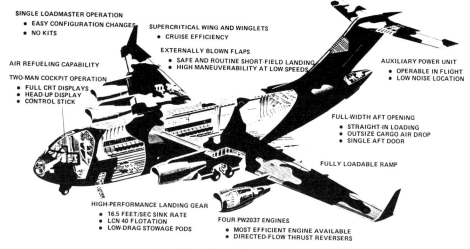

SINGLE LOADMASTER OPERATION
- EASY CONFIGURATION CHANGES
- NO KITS

SUPERCRITICAL WING AND WINGLETS
- CRUISE EFFICIENCY

EXTERNALLY BLOWN FLAPS
- SAFE AND ROUTINE SHORT-FIELD LANDING
- HIGH MANEUVERABILITY AT LOW SPEEDS

AIR REFUELING CAPABILITY

AUXILIARY POWER UNIT
- OPERABLE IN FLIGHT
- LOW NOISE LOCATION

TWO-MAN COCKPIT OPERATION
- FULL CRT DISPLAYS
- HEAD-UP DISPLAY
- CONTROL STICK

FULL-WIDTH AFT OPENING
- STRAIGHT-IN LOADING
- OUTSIZE CARGO AIR DROP
- SINGLE AFT DOOR

FULLY LOADABLE RAMP

HIGH-PERFORMANCE LANDING GEAR
- 16.5 FEET/SEC SINK RATE
- LCN 40 FLOTATION
- LOW-DRAG STOWAGE PODS

FOUR PW2037 ENGINES
- MOST EFFICIENT ENGINE AVAILABLE
- DIRECTED-FLOW THRUST REVERSERS

FIG 8.12.   Design features of C-17.

contracts to three US aerospace companies (Boeing, McDonnell Douglas and Lockheed) to study advanced tactical transport concepts.

Lockheed High Technology Test Bed (HTTB)

Conscious of the need to develop a replacement for the C-130 in the not-too-distant future and understandably intent upon maintaining its position as one of the world's leading manufacturers of military airlifters, Lockheed has for some time been conducting its own in-house research into future transport systems. With financial assistance from its network of suppliers and sub-contractors, plus the recent injection of government funds, Lockheed has now intensified its research and development programme, using a modified L-100-20 (the civilian version of the stretched C-130) as an airborne test platform. This aircraft has been designated the High Technology Test Bed (HTTB).

The aim of the HTTB programme is to study the feasibility of applying various new technologies—including aerodynamic modifications, high-lift devices, advanced composites, radar-absorbent materials, improved fly-by-wire flight controls and the latest avionics—to the production of a rugged, medium-sized, tactical airlifter capable of STOL operations from unpaved surfaces. Although the test schedule is still in its early stages, the initial trials have been encouraging. One notable innovation, designed to assist all-weather operations at unsophisticated forward airstrips, is the provision of a laser ranger and FLIR sensor in an under-nose turret controlled by a small joystick at the navigator's station. On approach, the laser is fired to determine the exact range to touch-down which, together with other essential information from the various flight instruments, is fed into the pilot's HUD to provide extremely accurate guidance to the predetermined landing point. At the same time, the FLIR display of the airstrip and runway can also be superimposed on the HUD to enable approaches without using ground-based landing aids.

Other modifications include the addition of a large dorsal strake in front of the fin and horsal strakes forward of the tailplane which have proved effective in straightening out the airflow at the rear of the aircraft and improving control at low speeds. Short field performance has been significantly enhanced by the provision of leading-edge droops (to increase the coefficient of lift), fast-acting double-slotted flaps, spoilers, and extended-chord rudder and ailerons. The HTTB has also been equipped with a high sink-rate landing gear that will enable the aircraft to land at more than one and a half times its previous sink-rate. Using specially designed shock absorbers, the new gear allows the aircraft to absorb a sink-rate of up to 14.7 ft/sec at a gross weight of 130,000 lb compared with the standard C-130's maximum sink-rate of 9 ft/sec at this weight. With these enhancements (indicated, with other major modifications, in Figure 8.13) the HTTB should be able to make a steep 6° approach at 75–80 kt (depending on weight) for a no-flare landing on rough surfaces followed by a ground run of under 1,500 ft.

Future Large Aircraft (FLA)

Meanwhile, under the auspices of the Independent European Programme Group (IEPG), the major European members of NATO have been working to define and harmonise their own requirements for a Future Large Aircraft (FLA). Although intended primarily as a tactical transport with some inter-theatre capability, the FLA—as its all-embracing title suggests—is also perceived as a potential replacement for all of NATO's large aircraft including those employed on tanker, AEW and maritime patrol duties. Such an aircraft, it is argued, would not only bring much needed rationalisation and commonality to NATO's air forces but would also yield worthwhile savings in research and development costs since these would be shared among the participants.

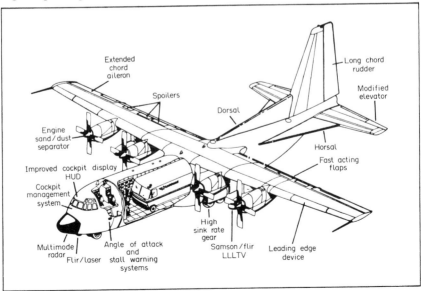

FIG 8.13.   Major modifications featured on Lockheed HTTB.

Preliminary studies are still at a very early stage, especially with respect to the feasibility of using a common airframe for such a wide variety of applications. Nevertheless, a consensus is now emerging on the characteristics required of the FLA airlifter. This is the version which is expected to command the largest market as countries such as Belgium, France, the Federal Republic of Germany, Italy and the UK face up to the need to select a replacement for their respective fleets of C-130s and C-160s. Most of these aircraft will reach the end of their operational life in the first decade of the next century and it clearly makes sense at least to consider whether their replacement can be produced as a collaborative venture. The next step in this process is to construct both an overall concept and a summary of the key operational criteria. Since these have yet to be announced, the characteristics described below represent merely the author's estimation of the capabilities likely to be required of the FLA.

Characteristics Required of the FLA Transport Variant

General Concept. The FLA transport variant must be a robust airlifter incorporating new but proven technology and capable of worldwide, all-weather operations in both the air-drop and air-land roles. Although intended primarily as a STOL-capable tactical transport, it must also be able to carry its maximum payload over distances of 1,800–2,000 nm, extendable by in-flight refuelling. The aircraft must be highly manoeuvrable on the ground and able to operate for long periods away from its main base with minimum support and maintenance. It will also be required to achieve better standards of survivability and reliability than the aircraft which it will replace.

Specific Requirements. Within the broad conceptual framework outlined above, it will of course be necessary for the states concerned to accept a degree of compromise between capabilities and costs, as well as between their differing national requirements. Specific characteristics to be resolved will include:

- Number and type of engines.
- Payload (in the region of 50,000–60,000 lb or 100 combat-equipped paratroops).
- Range with maximum payload in high-level cruise (1,800–2,000 nm).
- Radius of action for tactical (Hi-Lo-Hi) mission carrying 100 paratroops on outbound leg (1,500 nm including a total of 300 nm at low level).
- Cruise speeds (about 0.72 Mach at high level and 270–300 kt at low level).
- Field performance (take-off at maximum weight from runway not exceeding 6,000 ft and STOL landing on runway not exceeding 2,500 ft).
- Landing gear (number and type of wheels).
- Ground manoeuvrability (same as C-17).
- Cargo compartment (wide-bodied rectangular hold; rear ramp with same load-bearing strength as main floor; cargo handling system with quick-change capability; capacity for 10–12 standard pallets and medium-sized vehicles in double rows).

- Mission management aids (state-of-the-art avionics, including INS, EFIS and HUD; systems monitoring and fault diagnosis; station-keeping equipment; FLIR and other sensors for all-weather operations; all-round NVG compatibility).
- Crew (two pilots and one loadmaster).

Future International Military Airlifter (FIMA)

In parallel with and linked to the IEPG initiative to define and select a multi-role FLA, an international group of aerospace companies has been collaborating since 1982 on a project to design and develop a Future International Military Airlifter (FIMA). The four companies involved are Aérospatiale, British Aerospace, Lockheed and Messerschmitt-Bolkow-Blöhm. Since one of the main aims of this consortium is to have FIMA adopted as the FLA, it is not surprising that the two projects bear such a striking resemblance. Like the FLA, the embryonic FIMA project has yet to produce a detailed specification of the required performance criteria, but such information as is available suggests that the FIMA is virtually identical to the FLA in terms of basic goals and concepts.

For example, the FIMA is chiefly intended as a replacement for the C-130 and C-160 within the same timescale as that envisaged for the FLA though, like the FLA, it may also provide the basis for variants which can be employed in other roles. In general terms, the FIMA is intended to do roughly the same job as the C-130 but

PLATE 8.5. Artist's impression of FIMA aircraft engaged in an international airlift operation circa the year 2000.

in a more exacting environment and with better all-round performance predicated on the need to carry key items of large but not outsize military equipment into short airstrips close to the operational zones. In order to achieve this, the FIMA will need to strike a much better balance between its weight-lifting and bulk-carrying capabilities than the present generation of tactical airlifters. Other specific requirements will probably be much the same as those outlined earlier for the FLA, although there is likely to be a great deal of further horse-trading before both the countries and companies concerned agree a final design.

As with the C-17, albeit on a less expensive and ambitious scale, the FIMA concept essentially depends on the successful integration of several advanced aerospace technologies. This process should enable the production of an airlifter that will be significantly more capable than the aircraft which it replaces, and which should have wider applications than the tactical role for which it is optimised and mainly intended. The chances of attaining these objectives must be reasonably good, since all four partners in the consortium have extensive experience in manufacturing transport aircraft, especially Lockheed which will also be able to draw heavily upon its research and development programme with the HTTB. Although the FIMA is still an abstract concept, there is every reason to believe that it will be flying towards the end of the 1990s, provided the necessary funds can be made available.

PLATE 8.6. Artist's impression of FIMA in air-drop role. Note AAR probe and counter-rotating propellers.

V-22 Osprey

By now, it will have become apparent that one of the key elements in any future concept of air transport operations will be the ability to deliver substantial payloads into small, austere airfields. Ideally, the next generation of tactical airlifters should be able to operate in the Vertical/Short Take-Off and Landing (V/STOL) mode into the smallest airstrips or even into any clear area (such as a field, road or factory car park) large enough to accept the aircraft's wing span. However, while all of the new airlifters examined earlier in this Chapter will have a much better short-field performance than the aircraft they replace, none is designed for V/STOL operations. As a result, there will still be many occasions when troops and equipment will have considerable distances to cover, either by surface transport or helicopter, from the reception airfield to their ultimate destination. While helicopters will hence continue to play a very important role in providing logistic support and battlefield mobility for ground forces, their inherent disadvantages in terms of range, payload and speed in comparison with fixed-wing airlifters will continue to pose operational problems. In an attempt to overcome these drawbacks, a number of aerospace companies have been working for some time to design a new type of aircraft—a hybrid which will combine the agility, flexibility and vertical landing capability of the helicopter with the superior cruising levels, speed and range of the fixed-wing turboprop airlifter. The most notable achievement in this field to date is the development of the Bell-Boeing V-22 Osprey which was scheduled to make its first flight towards the end of 1988 and to enter service with all four US armed forces in 1992.

The V-22, which has been under development since 1981, works on the tilt-rotor principle. A streamlined nacelle on each wing tip houses a turboshaft engine, gearbox and rotor head to which is attached a three-bladed prop-rotor (so called because it also acts as a propeller when rotated forward). In the horizontal position, the rotors allow the V-22 to hover and manoeuvre, and take off and land vertically, like any other helicopter; but once airborne the entire nacelle (engine, transmission and rotor system) can be tilted through 90° by an actuator to transform the V-22 into a turboprop aircraft capable of speeds in excess of 275 kt. In the helicopter mode, the aircraft is controlled by varying the cyclic-pitch angle of each rotor; this alters the rotor disc's lift distribution thereby allowing asymmetric forces to move the aircraft forward, backward or to each side. The V-22 can accelerate in this mode through 100 kt in forward flight. Control in the cruise mode is maintained by use of flaperons, elevators and rudders. Flaperons are hinged control surfaces on the trailing edge of each wing which induce roll when acting as ailerons and provide increased lift when functioning as flaps. As on conventional aircraft, elevators and rudders on the tailplane are respectively used to control pitch and yaw. When the V-22's rotors are tilted forward as little as 20°, the aircraft becomes capable of STOL operations from any suitable clear surface. A rolling take-off in this configuration significantly increases the aircraft's payload capacity by increasing its maximum gross weight to 60,500 lb from the limit of 47,500 lb imposed for VTOL operations.

V-22 Mission and Capabilities

The V-22 is a multi-mission aircraft which has been ordered by all four Services in the USA. The USN plans to use it for anti-submarine warfare, logistic support and

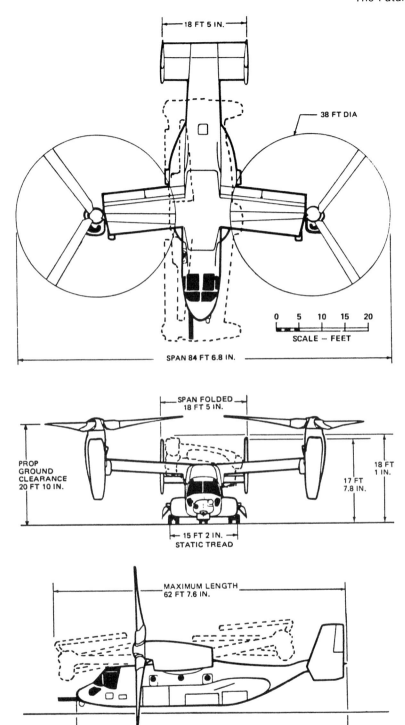

FIG 8.14.   Main dimensions of V-22 Osprey.

PLATE 8.7.   V-22 Ospreys engaged in assault operation.

search and rescue; the US Marine Corps and Army intend to use it mainly as a troop
and equipment airlifter, including the assault role; while the USAF plans to employ
it on deep-penetration special forces missions. Like the C-17, the design of the V-22
has been driven by the operational tasks which it is required to undertake. For
example, the USAF's requirement to hover at the mid-point of its long-range mission
established the V-22's total fuel capacity and engine power; the need to accommodate
24 combat-equipped troops determined the cabin area; and the requirement to carry

replacement engines for the Marine Corps F/A-18 Hornets governed the height of the cabin. Based on key criteria arising from the various operational requirements, the V-22 will be able to:

- Carry 24 combat-equipped troops or 12 stretcher patients over a radius of action of 200 nm.
- Carry 12 special forces troops over a radius of action of 520 nm and hover out of ground effect at the mid-point of the sortie at 4,000 ft in an ambient temperature of 35°C.
- Rescue four people within a 460 nm radius of action.
- Carry an internal payload of 20,000 lb.
- Carry an external payload of 15,000 lb on a dual-hook system.
- Cruise at speeds in excess of 275 kt.

To enhance these capabilities, the V-22 has been equipped with:

- Full-width rear door and ramp.
- Auxiliary power unit for self-sustainability.
- Multi-mode radar and night vision system.
- NVG-compatible cockpit.

The V-22 has a rectangular cabin of constant cross-section which can be entered either through side doors or via the full-width rear cargo ramp (see Plate 8.8). The designers have kept the cabin free from obstructions by employing an overhead wing

PLATE 8.8.   Artist's impression of V-22 engaged in special forces operation involving rescue of hostages.

and by using sponsons (similar to those on the CH-47D Chinook) to house the main landing gear and self-sealing fuel tanks. Fuel is also carried in self-sealing cells in each wing and there is provision for the aircraft to refuel in flight using a nose-mounted probe. With ferry tanks installed, the aircraft has an unrefuelled range of 2,100 nm making it highly self-deployable between as well as within theatres.

Advanced Features

The V-22 is the first US aircraft to have relied exclusively on computer-aided design, integrated with computer-aided manufacturing processes to improve production quality and fidelity. The V-22 has also been subjected to the largest programme of wind tunnel tests ever conducted for a rotary-wing aircraft. In all, 10 different scale models have undergone more than 8,000 hours of tests in eight wind tunnels. This was necessary not simply because it was essential to confirm the integrity of the revolutionary tilt-rotor concept but also due to the need to ensure that various other advanced features would perform as intended. For example, the airframe is built almost exclusively of composite materials, mainly in a solid-laminate structure of graphite epoxy. Composite structures are immensely strong yet weigh about 25% less than metal equivalents; they also offer stubborn resistance to corrosion, and are easy to repair. Its largely composite airframe and buoyant fuel sponsons also help to improve the V-22's flotation characteristics in the event of ditching. Finally, to ease crew workload, the V-22 has been equipped with digital flight management and fly-by-wire systems, fully integrated with the latest avionics and inertial navigation equipment.

Engines

The V-22 is equipped with the Allison T406-AD-400 engine, a derivative of the well proven T56 series which has been used for many years on the C-130. Indeed, the T56 was introduced in the late 1950s, since when it has been continuously updated with new technology, accumulating more than 130m flying hours in the process. The V-22's engines are 'two-shaft free turbines'. This means that the gas generator is not mechanically connected to the power turbine, thereby allowing the engines to be started and run up without turning the rotors—a useful option for the pilot in some situations.

Emergencies

The two engines are linked to an interconnecting drive shaft which enables the pilot to maintain balanced thrust with one power unit inoperative. This shaft automatically engages when the torque for one engine falls below a predetermined level, transmitting power from the good engine to the inoperative or failing engine's transmission and rotor system. Another feature provided to avert catastrophe is a safety mechanism which allows the rotors to revert to the helicopter mode to permit a vertical or run-on landing, should there be a failure of the actuator which operates the tilt system.

Summary of Leading Particulars

Crew: Two pilots and one crewman.
External dimensions: See Figure 8.14.
Internal dimensions:

	Cabin length	24.2 ft (7.37 m)
	Cabin width	5.9 ft (1.80 m)
	Cabin height	6.0 ft (1.83 m)
	Cabin volume	858 cu ft (24 m³).

Engines: Two Allison T406-AD-400 turboshafts each delivering 6,150 eshp.
Basic weight: 31,772 lb (14,411 kg).
Cargo hook capacity:

	Single	10,000 lb (4,536 kg)
	Dual	15,000 lb (6,804 kg).

Troop capacity: 24 (or 12 stretcher patients).
Maximum payload:

	Internal	20,000 lb (9,072 kg)
	External	15,000 lb (6,804 kg).

Take-off weights:

	Normal	42,000 lb (19,050 kg)
	Maximum (VTOL)	47,500 lb (21,545 kg)
	Maximum (STOL)	60,500 lb (27,442 kg).

Radius of action with 24 troops: 200 nm.
Self-deployment range: 2,100 nm.
Typical cruise speed: 275 kt.
Maximum cruise altitude: 27,000 ft.

CONCLUSION

Although transport operations do not enjoy the glamorous image of the offensive support, strike/attack and other combat-orientated roles, they share the classic air power attributes of speed, reach and flexibility. Moreover, they confer similar qualities upon the forces which they carry and support. Hence it is perhaps not surprising that, within the space of only a few decades, air transport operations have come to occupy a key role in the projection of military power, whether at the strategic, theatre or battlefield level.

This is not to suggest that airlift is the only important element in the force projection equation. On the contrary, surface movement (especially sealift) and the pre-positioning of equipment and supplies also have a significant part to play in the direct and indirect application of political and military power. Furthermore, military airlift is an expensive and relatively scarce resource in most countries. It must also be acknowledged that there are limitations on what air transport operations can achieve due to such factors as aircraft capacity, performance considerations and the availability of airfields in or adjacent to the operational areas in question. Nevertheless, only airlift can provide the rapid response which is so often essential if forces are to be inserted into a situation in sufficient time to deter, counter or defeat an attack.

With the advent of a new generation of considerably more capable and versatile airlifters, and with senior commanders likely to place increasing emphasis on air mobility in developing future concepts of operations, its seems certain that air transport forces will remain an indispensable instrument of national defence policy and power projection. Similarly, it is clear that the operational posture, reflexes and credibility of any national or international force will continue to be significantly enhanced by the extent to which its key components can be airlifted.

Questions

1. (a) Give some examples of 'outsize' loads.

   (b) Which aircraft in current service are able to carry 'outsize' loads?

2. Outline five of the main problems currently facing major air transport forces.

3. Why is the trans-shipment of loads from inter to intra-theatre aircraft so undesirable?

4. What broad principles are likely to influence future air transport operations?

5. List five important factors which governed the design of the McDonnell Douglas C-17.

6. Explain the principle of 'powered lift'.

7. Describe two advantages of the C-17's thrust-reverser system.

8. Identify four modifications used by Lockheed to improve the short field performance of the HTTB.

9. (a) Explain the 'tilt-rotor' principle used on the Bell-Boeing V-22 Osprey.

   (b) What provision is made on the V-22 for retention of balanced thrust in the event of loss of power from one engine?

Notes

CHAPTER 1

1. The first major airlift of troops is believed to have taken place in 1921 when, in rapid response to a major Kurdish rebellion, the RAF used Vickers Vernon transports to reinforce the British garrison at Kirkuk in Iraq. During further operations in Iraq and other areas of the Middle East in the 1920s, the RAF often used transport aircraft both for the speedy deployment of troops into actual or potential trouble spots and for the evacuation of casualties from locations which were otherwise inaccessible or very difficult to reach overland. The RAF also made effective use of transports in Afghanistan during the winter of 1928–29 to conduct what was probably the first ever air evacuation of civilians. With Kabul cut off and surrounded by rebel forces, the RAF successfully flew 586 men, women and children of various nationalities to safety over a period of two months.
2. W.F. Craven and J. L. Cate: *The Army Air Forces in World War II*. Vol VII p 19.
3. This quotation is cited by Wing Commander J. D. Brett on page 66 of his excellent booklet *A Short History of the Royal Air Force* published by the UK Ministry of Defence in 1984.
4. USAF, RAF, FAF and British civil aircraft collectively airlifted a total of 2,325,800 short tons into West Berlin, of which 1,586,000 tons were coal and 538,000 tons were food.
5. Established in 1983, France's *Force d'Action Rapide* contains 47,000 troops organised into five combat divisions. One of these (the 14,000-strong 11th Parachute Division, equipped with light armour) is usually at 24 hours' readiness for contingency operations outside Europe. However, in view of the French Air Force's limited airlift assets, a significant part of this division would either have to move by allied transport aircraft or by commercial airline if deployment was to be achieved within days rather than weeks.
6. Commanded by a 4-star general, US Cent Com includes five army divisions, a marine amphibious force, seven tactical fighter wings, two strategic bomber squadrons and substantial naval forces.
7. Interview with General George B. Crist, CinC US Central Command, *Marine Corps Gazette*, December 1986.

CHAPTER 2

1. The *Voyenno-Transportnaya Aviatsiya* (*VTA*) is the USSR's Military Transport Aviation Command—broadly the equivalent of the USA's Military Airlift Command (MAC).

2. In addition to MAC's 270 C-141 Bs, four C-141As are operated by the USAF's Aeronautical Systems Division and one C-141A is in service with NASA.
3. Other versions of the IL-76 are employed exclusively in the tanker and AEW roles.

CHAPTER 3

1. Like the USAF's KC-135s, the C-135F tankers operated by the French Air Force are also fitted with the boom system, but in practice a drogue adaptor is permanently attached to the end of the boom to permit refuelling by probe-equipped receiver aircraft.
2. See performance planning section in Chapter 5.
3. OMEGA is a long-range navigation aid based on a chain of eight VLF transmitters positioned at appropriate intervals around the globe. It enables suitably equipped aircraft to determine their position with great accuracy at all altitudes down to sea level.

CHAPTER 4

1. See Chapter 5 for description of VOR and TACAN.
2. See Note 3.3.
3. The People's Republic of China is still producing its own version of the AN-12, designated the Yunshuji-8 (transport aircraft 8) or Y-8. About 30 have been built to date. The aircraft is very similar, though not identical, to the Soviet version and is used mainly in the tactical transport role. Like the AN-12, the Y-8's cargo hold is currently unpressurised but the Chinese are thought to be developing a fully pressurised version that will be able to carry 100 troops at normal cruising altitudes.

CHAPTER 5

1. See Chapter 1, Figure 1.5.
2. When operationally justified, a reduction in net performance safety margins may be authorised in order to maximise airlift potential. Application of these reduced (and pre-calculated) parameters—sometimes referred to as 'military operating standards'—means, in effect, that aircraft can operate at heavier weights and hence carry heavier payloads.
3. Air which has been compressed in an engine compressor for use in pressurisation and other aircraft systems.
4. In crisis or war, commanders may authorise a reduction in fuel reserves, including diversion fuel, provided that the weather at the destination is forecast to be within prescribed limits at the time an aircraft is due to land. This measure allows an increase in payload at the expense of fuel, or may be taken to reduce the uplift of fuel at staging airfields.
5. RAF aircraft carry 'island holding' fuel when operating into Ascension Island (see Figure 1.2).

6. The commander of an aircraft carrying holding as opposed to diversion fuel will check and update the destination weather even more meticulously than usual, and will not proceed beyond his 'point of no return' (ie, that position along track beyond which an aircraft cannot safely return to its airfield of departure) unless the forecast weather at the destination is still favourable.

7. Just as senior commanders may authorise military operating standards in calculating aircraft performance (see Note 5.2) or reductions in fuel reserves (see Note 5.4) during war or other operational scenarios, so they may waive or extend normal crew duty criteria as circumstances demand.

CHAPTER 6

1. All aircrew employed in the tactical transport role need to develop a high level of proficiency, both as individuals in their respective specialisations and collectively as a crew. The exacting nature of the air-drop role (with its emphasis on low-level missions by day and night, precise navigation and accurate delivery of paratroops and supplies) calls for additional skills which can be achieved and maintained only through a regular and rigorous training cycle.

2. Overall losses of Allied aircraft during Operation MARKET GARDEN in The Netherlands in 1944 were relatively light, mainly because the Allies enjoyed a degree of air superiority which at times verged on total air supremacy. On the first day of the Operation (17 September 1944) nearly 1,500 fighters and fighter-bombers were used to escort the assault aircraft and suppress enemy flak batteries. Of the 1,024 C-47s and 120 gliders used to deliver the US 82nd and 101st Airborne Divisions onto their respective DZs near Nijmegen and Eindhoven, only 35 C-47s and 16 gliders were lost, mostly victims of flak; and of the 279 C-47s, 240 converted bombers and 320 gliders used to deliver the British 1st Airborne Division onto its DZs and landing zones near Arnhem during that same first afternoon of the Operation, not one was lost to enemy action, although 12 gliders ditched or force-landed short of their objectives owing to technical problems. See *Airborne to Battle* (*A History of Airborne Warfare* 1918–1971) by Maurice Tugwell.

3. Suppression of Enemy Air Defences (SEAD) is a specialist combat role usually undertaken in support of interdiction or strike/attack missions. SEAD resources are scarce and unlikely to be available to support an air-drop operation.

CHAPTER 7

1. Published annually by the International Institute for Strategic Studies, *The Military Balance* provides an excellent quantitative assessment of the military forces and defence expenditures of over 140 countries.

2. Nap-of-the-earth is a term used in helicopter operations to describe the technique of flying at very low level in order to take maximum advantage of terrain screening and features such as woods which in turn minimises exposure to visual and sensor acquisition.

3. See Volume 9 in this series.

4. *Military Helicopters*, P. G. Harrison *et al*, Brassey's, 1985, p. 8. Readers wishing to study helicopter survivability in more depth are recommended to read Chapter 7 of this book.

CHAPTER 8

1. Aircraft available for military use from the USA's Civil Reserve Air Fleet (CRAF) fall into two categories: those procured new with built-in cargo convertibility features, and other wide-bodied airliners already in commercial service which have been retrofitted to make them suitable for military airlift tasks. A recent programme to enhance the CRAF has concentrated on expanding the number of retrofitted aircraft as a means of easing the shortage of airlift for contingency operations.

Answers to Self-Test Questions

CHAPTER 2 – STRATEGIC OPERATIONS

1. Large payload in terms of both weight and size. Ability to carry personnel and/or cargo. Long range (at least 2,500 nm with maximum payload). Rapid on-load and off-load facilities. High cruising speed (at least 0.75 mach).
2. (a) Strategic freighter.
   (b) Flying boom.
   (c) To detect, identify, analyse and record systems malfunctions.
3. This allows vehicles to drive through the entire length of the cargo compartment, loading via one entrance and off-loading via the other without any need for reversing, and thereby affording a valuable reduction in turn-round time.
4. 'Bulk out' occurs when the volumetric capacity of a cargo hold is used up before reaching the aircraft's maximum payload limit in terms of weight. In the case of the C-141, the problem was alleviated by lengthening the fuselage by 280 ins which increased the volumetric capacity of the cargo hold by some 25%, thereby allowing greater exploitation of the aircraft's weight-lifting potential.
5. When it is necessary to mount air-drop operations over strategic distances.
6. The AN-124 is larger than the C-5B and can carry more payload and fuel. However, unlike the C-5B, the AN-124 is equipped neither for in-flight refuelling nor for air-drop operations and is hence less versatile.
7. This policy confers great flexibility and interchangeability, allowing Aeroflot aircraft to be used on routine military missions when so required, or on special military tasks when the use of *VTA* aircraft might be too provocative or attract too much attention. Aeroflot's large fleets of variants of military airlifters also provide a valuable and immediately available transport reserve, should the *VTA*'s resources prove insufficient in a time of national crisis.
8. (a) This configuration poses structural problems by inducing drag and heavy aerodynamic loads on the rear fuselage. The designers of the AN-22 overcame this by equipping the aircraft with a twin-fin tail topped off with anti-flutter devices.
   (b) By exploiting the principle of 'blown flap'.

CHAPTER 3 – TANKER/TRANSPORT OPERATIONS

1. (a) Such an aircraft must be able to undertake both its tanker and transport roles simultaneously (ie, during the same mission).
   (b) This is a mission during which a tanker/transport provides AAR for combat aircraft while simultaneously airlifting their support personnel and equipment.
2. Flight refuelling can be highly effective in enhancing the reach, scope and flexibility of airlift operations by allowing AAR-equipped transport aircraft to

carry heavier payloads in shorter timescales over longer ranges with less dependence on intermediate staging posts.

3. (a) 'Flying boom' and 'probe and drogue'.

   (b) The flying boom system permits a higher flow-rate but fuel can be transferred only to one receiver at a time, whereas hose and drogue tankers can dispense fuel simultaneously to three receivers. The flying boom system requires a specialist boom operator who is responsible for 'flying' the boom into the receiver's socket, but the receiving pilot must maintain a very stable profile throughout the transfer process. By contrast, while the pilot of a probe-equipped receiver is responsible for the initial engagement, he has more room for manoeuvre during the transfer process thanks to the inherent flexibility of the dispensing hose.

4. This technique may be used when a particular combination of tanker and receiver (eg, a Tristar and Hercules) leads to the receiver being subjected to strong downwash during the transfer process. After commencing AAR at, say, 25,000 ft both aircraft gradually descend to about 8,000 ft while fuel is being transferred, thereby allowing the receiver to accept an increasing all-up weight without having to apply climb power just to maintain straight and level flight.

5. This could be crucial to maintaining the momentum of an airlift and/or AAR operation, especially when either the tanker/transport itself or the aircraft it is supporting are required to cover long sectors with no staging facilities *en route*. The ability to receive as well as dispense fuel could also be used to overcome restrictions in take-off weight at the departure airfield by allowing a tanker/transport to become airborne with maximum payload and top up its tanks *en route*.

6. The KC-10 has a longer boom which permits greater separation during refuelling. The KC-10's boom is guided by a sophisticated fly-by-wire aerofoil which provides the excellent controllability needed for fast and accurate link-ups with receiver aircraft.

   The KC-10's boom system features an exceptionally high flow-rate (up to a maximum of 1,500 US gallons/min.).

7. Unlike the other Tristar variants, the KC Mk 1 is specially designed for the cargo role. To allow the airlift of large and heavy items of freight, the aircraft is equipped with a cargo door, strengthened floor and integral pallet-handling system.

8. 'Active aileron controls' automatically apply load-relieving inputs to the outboard ailerons when wing loading is increased due, for example, to manoeuvres and wind gusts. By reducing lift over the outer wings, this effectively redistributes total lift inboard, relieving flexing moments and stress on the overall wing structure.

CHAPTER 4 – TACTICAL OPERATIONS

1. Possession of a tactical airlift capability can enhance a state's military potential (out of all proportion to the resources involved) by conferring a degree of flexibility and mobility in the exertion of force that would otherwise be impossible to achieve.

2. Rugged construction. Good payload and well designed cargo compartment. Rear doors and ramp to facilitate cargo handling. Rapid reconfiguration from one role to another. Air-drop capability.

3. High wing; low-slung fuselage; upswept tail; rear doors and ramp; high-lift devices; heavy duty landing gear.

4. (a) By definition, such aircraft offer a combination of relatively large payload and endurance.

   (b) Tanker.
   Gunship.
   Minelayer.
   Search and rescue.
   Weather reconnaissance.

5. The AN-12's main cabin is unpressurised. This means that, if passengers are carried, the aircraft cannot fly above 10,000 ft thus reducing its range by increasing its fuel consumption. This problem is exacerbated by the AN-12's limited fuel capacity which is only half that of the C-130H. Finally, the AN-12 lacks an integral rear ramp.

6. As a result of acquiring an additional fuel tank in the centre section of its stengthened wing, the second-series C-160 has a maximum fuel load of 47,800 lb compared with the earlier variant's maximum of 32,500 lb. Moreover, the second-series aircraft is equipped for in-flight refuelling. It also has better avionics.

7. (a) Fokker F-27 Mk 400M.

   (b) Of the aircraft described in Chapter 4, only the C-160 has a true kneel capability. However, pressure in the G-222's oleopneumatic shock absorbers can be adjusted to vary the height and attitude of the cargo compartment floor in relation to the ground.

8. Tactical airlift can be indispensable to the maintenance of essential logistic support for air combat units. If the units are to be sustained at maximum strength and efficiency, they must receive a regular supply of high-priority spares, including replacement engines, even when their bases are disrupted by enemy action. Tactical transports such as the C-23A can provide such a service by virtue of their ability to operate from grass areas, taxiways or undamaged sections of runways at the bases used for air combat operations.

CHAPTER 5 – FACTORS INVOLVED IN PLANNING AIRLIFT OPERATIONS

1. Dimensions, condition and load-bearing strength of runways, taxiways and parking ramps.
   Availability of approach and landing aids, emergency services and ground handling facilities.

2. (a) It establishes a safe relationship between the strength of a paved surface expressed in terms of LCG and the weight of an aircraft expressed in terms of LCN.

   (b) It can use the airfield occasionally subject to the approval of the relevant authorities.

3. Aerial port, maintenance and communications.

4. (a) To ensure that the room required for a manoeuvre is never more than the space available.
   (b) Take-Off: from the beginning of the take-off run to an initial height of 35 ft. Take-Off Net Flight Path: from the initial height of 35 ft to a height of 1,500 ft above the airfield.
   En Route Stage: from 1,500 ft on departure to 1,500 ft on the approach to the destination.
   Landing: from 1,500 ft on the approach to the point on the runway where the aircraft completes its ground roll.
5. The WAT limit establishes the maximum aircraft weight for a particular combination of airfield altitude and temperature, ensuring that specified gradients of climb can be achieved with one power unit inoperative.
6. Selection of route.
   Availability of diversion airfields.
   Crew availability and utilisation.
   Capacity of reception airfield.
   Cargo off-load procedures.
7. Type of aircraft.
   Type and amount of cargo.
   Load configuration.
   Availability of handling equipment.
   Expertise of loadmaster and aerial port personnel (loading and unloading an aircraft requires specialist skills).
8. Advantages: EROs reduce turn-round time, ease parking ramp congestion and increase mission flow-rate. They also reduce the risk of being unable to re-start one or more engines. In hostile scenarios, they reduce aircraft vulnerability.
   Disadvantages: Can be hazardous to deplaning passengers or aerial port and maintenance personnel working in the vicinity unless all concerned are well briefed and preferably experienced in ERO procedures.

CHAPTER 6 – AIR-DROP OPERATIONS AND TECHNIQUES

1. Seizure of point of entry.
   Coup de main operation.
   Reconnoitre and report enemy activity.
   Attack command, control and communications facilities or other key points.
   Rescue or protect ex-patriates.
2. Mounting the operation by night.
   Flying as low as possible.
   Using 'tactical' routes.
   Maintaining radio and radar silence.
   Dropping personnel and equipment from the lowest practicable height.
   Operating in small elements, joining up only for the run-in to the DZ.
3. Should be a prominent geographic feature about 10 nm before the DZ.
   Track from TAP should run down the longest axis of the DZ.
   Must take due account of enemy dispositions.

4. Personnel or equipment leave the aircraft at high altitude and complete a free-fall phase before their parachutes are opened for the final descent.

5. Each of these systems enables equipment to be air-dropped with great precision.

6. (a) SKE enables a stream of tactical aircraft to establish and maintain a safe formation even when they lose all visual references to one another, thereby rendering air-drop operations much less susceptible to bad weather.

   (b) SKE emits a distinctive radar footprint and is highly vulnerable to jamming.

   (c) Apart from its ability to provide short-range distance and bearing from a ground beacon on the DZ (if available), SKE gives no positional information and does not of itself permit aircraft to navigate in formation along a predetermined track.

7. SKE and INS.

8. This is a predetermined datum about 4 nm before the DZ where aircraft initiate a climb from low level to their dropping height, while simultaneously reducing to dropping speed. Aircraft are at their most vulnerable from the pop-up point until they have completed their drop and can once again accelerate and descend to low level.

CHAPTER 7 – SUPPORT HELICOPTER OPERATIONS

1. The airmobile concept is based on the use of helicopters to provide increased mobility for ground combat forces on and around the battlefield, thereby enhancing the ground commander's ability to react quickly to a changing tactical situation over a wide area.

2. Tactical deployment of troops and weapons.
   Assaults and raids.
   Infiltration.
   Redeployment.

3. Size and shape of the load.
   Type of load.
   Time and distance.
   Cargo handling requirements.
   Freedom of manoeuvre.

4. Rapid deployment of security forces.
   Positioning and extraction of rooftop and surface observation teams.
   Aerial reconnaissance and surveillance.
   Reinforcement and resupply of garrisons which are inaccessible by road.
   Casualty evacuation.

5. Adoption of an ultra low level nap-of-the-earth profile, making full use of terrain screening.
   Use of passive and active countermeasures (including warning sensors, jammers, chaff and infra-red decoys).
   Protection and duplication of vital systems.

6. Reduces aircraft manoeuvrability and transit speed.
   Hinders use of terrain-screening flight profile.
   Induces extra drag which increases fuel burn and hence reduces range.

7. Extremely low temperatures may cause oils and other aircraft fluids to congeal. Snow blown up by the rotors may obscure visibility on take-off and landing. Severe storms and blizzards may bring all helicopter operations to a standstill. Featureless terrain can make navigation very difficult. Concealment may be very difficult to achieve.

8. The air plan must always be based upon and subordinate to the ground tactical plan.

9. As a result of its more powerful engines, the Mi-17 has a better performance than the Mi-8. This allows the Mi-17 to fly higher, faster and further – and carry some 2,000 lb more payload – than the Mi-8.

10. Tandem rotors cancel out each other's torque, thereby obviating the need for a tail rotor and permitting the entire length of the fuselage to be utilised for payload with no space wasted on an empty tailboom. They also improve handling in the hover and, by 'spreading' the aircraft's centre of gravity, allow greater flexibility in the disposition of both internal and external loads.

CHAPTER 8 – THE FUTURE

1. (a) Heavy trucks, self-propelled artillery, CH-47D helicopters, bulldozers and main battle tanks.
   (b) AN-22, AN-124 and C-5.

2. Shortfall in total capacity.
   Shortfall in outsize capacity.
   Shortage of reception airfields.
   Transloading from inter- to intra-theatre airlifters.
   Age of airlift fleets.

3. It is time-consuming, expensive and entails duplication of cargo handling equipment as well as additional personnel. In short, it is an inefficient way of running an airlift operation.

4. In future, airlift forces will be required to deliver all items of air-portable equipment directly into austere airfields as close as possible to their ultimate destination. It follows that the new airlifters acquired to undertake such missions will need better short-field performance, ground manoeuvrability, maintainability and all-round versatility than the aircraft they replace.

5. Size, weight and amount of combat equipment to be airlifted.
   Inter-theatre distances.
   Airfield dimensions.
   Ground manoeuvrability.
   Reliability.

6. 'Powered lift' is obtained by directing engine efflux onto and through large double-slotted flaps. By increasing the wings' coefficient of lift, this enables an aircraft to fly at much slower speeds than would otherwise be possible.

7. First, reverse thrust can be selected at any speed on any surface with minimal risk of disturbing and then ingesting harmful debris, thus facilitating landings on short, unpaved runways. Second, because efflux is vectored upward with the engines in reverse, EROs can be conducted with no risk to personnel working in the vicinity of the aircraft.

8. Leading-edge droops; fast acting double-slotted flaps; spoilers; extended-chord rudder and ailerons.

9. (a) In the horizontal position, the rotors allow the V-22 to take off and land vertically and hover like any other helicopter, but once airborne they are tilted forward progressively to allow a transition to high-speed flight when the rotors effectively act as large propellers.

   (b) When torque on one engine falls below a predetermined level, an interconnecting drive shaft automatically engages to transmit power from the remaining engine to the other's transmission and rotor system.

Index